Endocrinology Case Studies

Sharmin Jahan

Endocrinology Case Studies

Solving Complex Endocrine Disorders

Springer

Sharmin Jahan
Department of Endocrinology
Bangabandhu Sheikh Mujib Medical University
Dhaka, Bangladesh

ISBN 978-981-95-2769-4 ISBN 978-981-95-2770-0 (eBook)
https://doi.org/10.1007/978-981-95-2770-0

This Springer imprint is published by the registered company Springer Nature Singapore Pte Ltd.
The registered company address is: 152 Beach Road, #21-01/04 Gateway East, Singapore 189721, Singapore

To the loving memory of my parents
Their unwavering support and unshakeable faith in me
Made me the person I am today

Foreword

Although the history of endocrinology is variously traced back more than two thousand years to China and ancient Greece, the description of specific diseases and etiologies appeared in the nineteenth century through Berthold, Graves, von Basedow, and Addison, with Bayliss and Starling using the term "hormone" in the early twentieth century. The discipline of endocrinology came of age in the second half of the twentieth century with the development of robust measurement of, and chemical purification of, the hormones secreted by the organs that constitute the endocrine system. As endocrinology transitioned to a robust science, a clear understanding of the pathophysiology of endocrine diseases and therapeutic strategies emerged. Addison disease for instance went in a lifetime from an almost invariably fatal disease to a condition that responded to adrenal extract with dosage monitored by the eosinophil count, then further transformed by the isolation of cortisol with increasingly sophisticated technology for measurement and testing.

Modern textbooks of endocrinology now run to several volumes representing a knowledge base that can be challenging for all but the most experienced endocrinologist. Despite the clarity to be found in these texts, life is inevitably more complicated and after many years of practice one wonders whether atypical presentations of endocrine conditions are more "classical" than the well-documented "classical presentations."

In this volume, Professor Sharmin Jahan explores the diversity encountered in endocrine diagnosis through cases encountered at her institution covering diseases of the adrenal, parathyroid, pituitary, testis, and thyroid glands involving disordered aldosterone, androgen, calcitonin, catecholamine, chorionic gonadotropin (hCG), cortisol, growth hormone, parathyroid hormone, parathyroid-related peptide, and testosterone secretion. The patients had presented with a range of symptoms and signs, not always classically associated with endocrine disease; they occurred on the basis of genetics, neoplasia, and autoimmunity.

This case series is a welcome addition to the endocrine literature coming as it does from Dhaka in Bangladesh. South Asian centers are generally underrepresented in the clinical literature, and although most if not all the conditions described are not unique to any ethnicity or geography, critical nuances in symptoms and signs emerge that need to recognized and understood if patients of any background are to receive both appropriate and holistic endocrine treatment. This should be a welcome addition to any library particularly for those entering or early-in their endocrinology careers.

Peter J. Fuller
peter.fuller@hudson.org.au
http://hudson.org.au/

Centre for Endocrinology and Reproductive Health
Hudson Institute of Medical Research
Monash Medical Centre
Clayton, Australia

Preface

I have always been fascinated by the puzzle of difficult endocrine cases. Dealing with a perplexing endocrine case presented with a seemingly incoherent set of clinical data made further difficult with apparently unrelated investigations looked like a daunting task. It always gave me immense pleasure to go through the case thoughtfully to unveil the final diagnosis. As the renowned endocrinologist Prof. William F. Young Jr. once described, "*Being a good endocrinologist is like being a good detective having a sharp clinical eye to solve the mystery of enigmatic endocrine cases.*"

I have been working as an endocrinologist for the last 15 years or so. As part of my job, I had to conduct the weekly clinical case presentation of our department. Also, preparing and presenting unique endocrine cases for presentation at any national/international conference seemed quite challenging but very rewarding at the end. Scrutinizing each case made it clear that consulting the literature only provided surface knowledge; we realized that *"A single patient is better than a thousand books".*

During the last three years, I have started compiling the presentations to keep a permanent document of the wonderful cases and decided to write a book to discuss the rational approach in dealing with such a case. The cases have been written in a descriptive manner with discussion on relevant areas in MCQ format. The answers are provided at the end of each case along with a brief discussion on it, strengthened by relevant review literature. Some of the endocrine cases are relatively straightforward while others are unique in presentation and diagnosis. I have deliberately kept an admixture of such cases so that it is palatable to both the endocrine students and endocrinologists as well as physicians in other specialties.

This is my first attempt at writing such a case-based book. I humbly request that you forgive any mistakes committed unwillingly. I shall also highly appreciate any comments or advice on how to improve it in the future.

Dhaka, Bangladesh Sharmin Jahan

Acknowledgments

The journey of publishing this book has been really challenging yet immensely rewarding. It would not have been possible without the guidance, support, and encouragement of so many individuals to whom I am deeply grateful.

First and foremost, I would like to thank my patients, whose stories—though anonymized and respectfully adapted—form the foundation of this book.

I am deeply indebted to my mentors, colleagues, and students in the field of endocrinology at BSMMU, whose wisdom and insights have shaped my clinical acumen and academic growth. Their engagement in the critical discussions of the cases has greatly enriched this work.

My special thanks go to my beloved students Dr. Tahseen Mahmood and Dr. Md. Fakhrul Alam, for their relentless endeavor in establishing communication with the publisher and helping in the editing process. Without their all-out effort, it would not have been possible to bring this book to light.

I also extend my sincere gratitude to the editorial and publishing team, whose professionalism, patience, and meticulous attention to details have been instrumental in transforming this manuscript into a polished publication.

I consider myself unmeasurably blessed to have the endless encouragement, understanding, and love of my husband. Your belief in me and this project kept me going through long hours and complex cases.

Finally, I would like to express my gratitude to my nearest kin whose continued inspiration turned into the major driving force behind this project during times of uncertainty.

This book is dedicated to the ongoing pursuit of medical understanding and to all those committed to improving the lives of patients with endocrine disorders.

Contents

About the Author

Sharmin Jahan passed MBBS from Dhaka Medical College in 2000 with distinction. She completed her fellowship in Internal Medicine from Bangladesh College of Physicians and Surgeons (BCPS) in 2008. She obtained MD in Endocrinology and Metabolism and received Prime Minister Gold Medal in 2014 for securing the highest marks among all the subjects in the Faculty of Medicine. At present, she is working as a Professor in the Department of Endocrinology, BSMMU, Dhaka, Bangladesh. She has presented various original works and endocrine-related papers at conferences at home and abroad. She has published about 77 original articles in esteemed international journals and a sizable number in national papers. She is the principal investigator in research projects on pituitary disorders, especially Cushing disease, growth, and pubertal disorders as well as adrenal disease including endocrine hypertension at her institute. She is also a co-investigator in the studies conducted on gestational DM and DM in the youth in BSMMU. In 2018 she had training at Mayo Clinic, Rochester, Minnesota, USA, under Prof. William F. Young, Chair, Division of Endocrinology, with emphasis on pituitary, adrenal, and gonadal disorders. Currently, she is in the final phase of her PhD program on "Endocrine hypertension" at Monash University, Melbourne, with fully funded scholarship under Prof. Peter Fuller and Dr. Jun Yang.

A 39-Year-Old Male With Vomiting, Diarrhea, and Abdominal Pain

1

Introduction

Hypercalcemia is a common clinical problem attributable to primary hyperparathyroidism or malignancy in most cases. Though most patients with primary hyperparathyroidism are asymptomatic, infrequently, it may present with frank complications or even with hypercalcemic crisis. The following case demonstrates the classic presentation of hypercalcemic crisis in a 39-year-old male and highlights the rational steps in evaluation with subsequent management.

Case Summary

A 39-year-old normotensive, nondiabetic male presented with vomiting associated with abdominal pain for the last 01 month. It was projectile in nature occurring six to seven times daily, and the vomitus contained food particles, neither bile nor blood-stained. It was associated with episodic, colicky epigastric pain, insidious in onset and aggravated on empty stomach, relieved by food, occasionally radiating to back. He consulted a physician and was investigated accordingly and found to have mild anemia with raised serum creatinine (Table 1.1). His USG of the whole abdomen showed bilateral nephrocalcinosis (Fig. 1.1). Further evaluation by a nephrologist revealed severe hypercalcemia and hypophosphatemia with mild renal impairment (Table 1.1).

S. Jahan, *Endocrinology Case Studies*,
https://doi.org/10.1007/978-981-95-2770-0_1

Table 1.1 Biochemical report of the patient showing severe hypercalcemia with renal impairment

Parameter	Patient's value	Reference value
CBC and ESR	Hb—12.5g/dL, ESR—10 mm, total WBC count—8200/cu mm, platelet count—3,50000/cu mm	Hb—15±2 gm/dL, ESR—0–10 mm in 1st h, TC of WBC—7±3 × 10^9/L Platelet count—150000–400000/cmm
SGPT	40 U/L	14–63 IU/L
serum amylase	90 IU/L	30–110 IU/L
Serum creatinine	**1.8 mg/dL (1st occasion)** **1.9 mg/dL (2nd occasion)**	0.5–1.3 mg/dL
Serum calcium	**14.9 mg/dL**	8.5–10 mg/dL
Serum phosphate	**2.2 mg/dL**	3-4.5 mg/dL

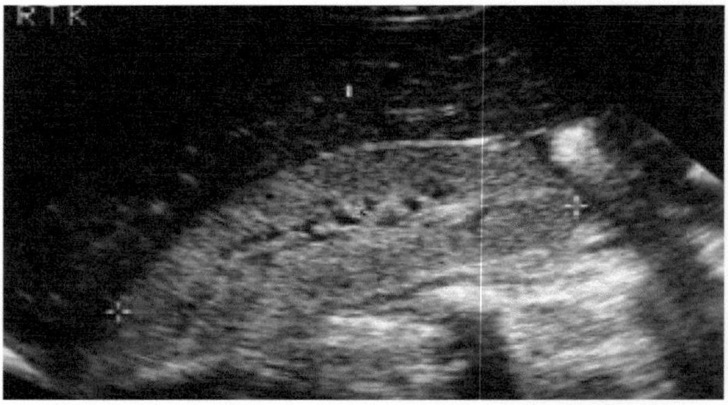

Fig. 1.1 USG of abdomen showing nephrocalcinosis

Q1. What are the possible causes of hypercalcemia in this patient?
  (a) Primary hyperparathyroidism
  (b) Secondary hyperparathyroidism
  (c) Tertiary hyperparathyroidism
  (d) Hypercalcemia of malignancy
  (e) Familial hypocalceuric hypercalcemia

Q2. Mention the probable reasons for his abdominal pain:
  (a) Peptic ulcer disease
  (b) Nephrocalcinosis
  (c) Acute pancreatitis
  (d) All of the above
  (e) None of the above

The patient on query gave a history of generalized body ache, polyuria, nocturia, and polydipsia for the last 6 months. He increasingly became lethargic and fatigued during the course of illness. He gave no history suggestive of chronic kidney disease (CKD) or any underlying etiology contributing to CKD. He had no family history of a similar illness. However, he gave H/O recent weight loss of approx.10 kg. He was found to be mildly emaciated and moderately anemic and dehydrated, BP: 130/80 mm Hg, other vital parameters were normal, there was no thyromegaly, lymphadenopathy, bony tenderness, and systemic examination failed to reveal any other abnormality. So, at this point, the probable diagnosis was primary hyperparathyroidism as clinical clues to any underlying malignancy (smoking, hemoptysis, cough, chest pain, lymphadenopathy, etc.) or CKD (nocturia, oliguria, edema, DM, HTN, vasculitis, etc.) were absent. Biochemical abnormalities were also supportive of primary hyperparathyroidism like hypercalcemia and hypophosphatemia. Elevated serum creatinine probably reflected dehydration leading to pre-renal azotemia, and nephrocalcinosis was a contributory factor to renal impairment. His epigastric pain resulted from peptic ulcer disease probably precipitated by hypercalcemia-induced secretion of excess gastrin. Hypercalcemia is an important metabolic cause of acute pancreatitis leading to sudden, severe epigastric pain. It may also lead to loin pain occurring from nephrocalcinosis which might be associated with dysuria, hematuria, and renal impairment.

On admission to the endocrine ward, he was found to be drowsy and severely dehydrated; serum calcium was 18 mg/dL, indicative of a hypercalcemic crisis.

Q3. What should the initial management be?
 (a) I/V normal saline: 3–4 L/day
 (b) Inj. furosemide: 1 amp b.d
 (c) Inj. zoledronic acid
 (d) Recombinant PTH
 (e) Sevelamer

The patient was rapidly rehydrated with N/S infusion and unrestricted plain water intake, followed by saline diuresis with furosemide. The next day he received inj. zoledronic acid after confirmation of safe eGFR (38 mL/min/1.73 m^2). After volume resuscitation, his renal function improved and serum creatinine came down to 1.4 mg/dL, and serum calcium plateaued at 9.5 mg/dL. Recombinant PTH is used in the management of osteoporosis, and it is contraindicated in hypercalcemic crisis in primary hyperparathyroidism. Sevelamer is a phosphate binder used in 2°/3° hyperparathyroidism.

Q4. What should be the next plan of evaluation?
 (a) Plasma vitamin D level
 (b) Bone scan
 (c) SPECT-CT fusion image of the neck, chest/ T_{99} sestamibi scintigraphy of neck
 (d) USG of the neck
 (e) Plasma intact PTH level.

After the improvement of clinical conditions, he underwent further investigations. Plasma intact PTH level was remarkably high—565 ng/mL. USG of the neck revealed enlarged left lower parathyroid gland suspicious of parathyroid adenoma/hyperplasia. The finding was further confirmed by the SPECT-CT image of the neck (Fig. 1.2). USG of W/A showed nephrocalcinosis with the early renal parenchymal disease. Plasma vitamin D level was found to be normal

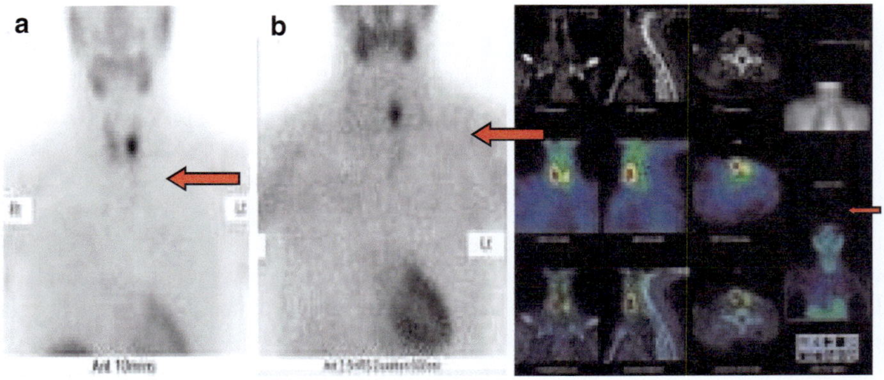

Fig. 1.2 T_{99} scintigraphy (**a** and **b** on left panel) and SPECT-CT (right panel) image of the neck showing a left lower parathyroid

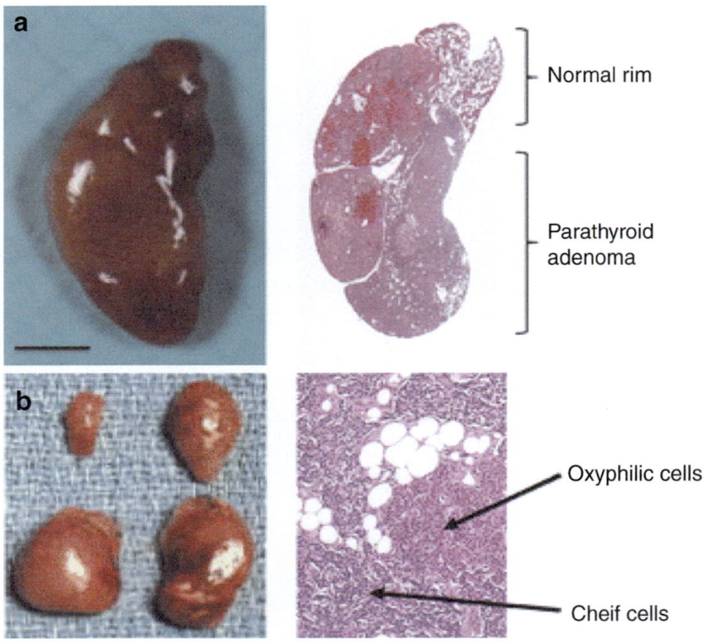

Fig. 1.3 The gross appearance (**a** and **b** on left panel) and histopathology (upper and lower image on right panel) of excised parathyroid adenoma. (Source: Barczyński et al. [2])

as expected excluding secondary hyperparathyroidism due to vitamin D deficiency. The bone scan is indicated to look for metastatic involvement in malignant hypercalcemia, a less likely diagnosis in this case. Later, he was referred to a parathyroid surgeon and underwent open parathyroidectomy. His histopathology report was suggestive of parathyroid adenoma. Currently, he is on remission having regular follow-up (Fig. 1.3).

Discussion

Hypercalcemia is a common metabolic problem encountered by the clinician. 90% of cases can be attributed to primary hyperparathyroidism or hypercalcemia of malignancy. Patients with primary hyperparathyroidism are asymptomatic in about 85% of cases. However, they can also present with body aches, bone pain, pathologic fracture, abdominal pain due to PUD, renal calculi or acute pancreatitis and depression, apathy, or lethargy reminiscent of the adage—*bones, stones, abdominal groans, psychic moans, and fatigue overtones*. At times, it may present with frank hypercalcemic crisis manifested by polyuria, polydipsia, dehydration, drowsiness, and a depressed level of consciousness that constitutes an endocrine emergency. 85% cases of primary hyperparathyroidism are attributable to single/multiple adenomas, 15% due to parathyroid hyperplasia, and <1% due to parathyroid carcinoma. It can be sporadic or occur as a component of a familial syndrome like MEN 1 or 2, hyperparathyroidism-jaw tumor syndrome (HPT-JT syndrome), isolated familial hyperparathyroidism, etc. Biochemically, it is characterized by high calcium, low/low normal phosphate, and high intact PTH. Normal plasma vitamin D, low phosphate, and normal/minimally elevated serum creatinine level in these patients exclude secondary/tertiary hyperparathyroidism due to CKD or vitamin D deficiency. Management of hypercalcemic crises include correction of dehydration with N/S, forced saline diuresis, I/V bisphosphonate, S/C calcitonin, glucocorticoids (specially in malignancy), or hemodialysis in recalcitrant cases. Specific management depends on the localization of parathyroid adenoma/hyperplasia by USG/CT/MRI of the neck or with parathyroid scintigraphy. At least two modalities of imaging are required to avoid false localization. However, localization is not mandatory for final diagnosis, and a high-volume parathyroid surgeon can localize the tumor during neck exploration in >80% of cases [1].

Answers

Q1. (b)
Q2. (a), (d)
Q3. (a)
Q4. (a), (c) (d), (e)

References

1. Bringhurst FR, et al. Hormones and disorders of mineral metabolism. In: Melmed S, et al., editors. William's textbook of endocrinology. 12th ed. Philadelphia: Elsevier Saunders; 2011. p. 1261–9.
2. Barczyński M, et al. Sporadic multiple parathyroid gland disease-a consensus report of the European Society of Endocrine Surgeons (ESES). Langenbecks Arch Surg. 2015;400:887. https://doi.org/10.1007/s00423-015-1348-1.

A 40-Year-Old Presented with Abdominal Fullness, Generalized Red Spots with Recent-Onset Diabetes and HTN

2

Introduction

A 40-year-old male presented with clinical and biochemical feature of Cushing syndrome. He had a large adrenal tumor co-secreting glucocorticoid and adrenal androgen, later diagnosed as having an adrenocortical carcinoma (ACC). The case scenario highlights the clinical, biochemical clues and the imaging phenotype of ACC and the differentials of a large adrenal incidentaloma. There is a brief discussion on clinical presentation, investigations, and management of such a tumor which often has a grave prognosis.

Case Scenario

A 40- year-old male got admitted with the complaints of abdominal fullness for the last 8 months. The symptom was not associated with any abdominal pain, nausea, vomiting, or altered bowel habit. He also noticed multiple, evolving, generalized, non-itchy, nontender red spots over the last $1^{1/2}$ months. On query, he gave H/O difficulty in standing from sitting position (1½ months) that improved spontaneously and was not related to the intake of any drugs/toxins. He also noticed bilateral leg swelling for the same duration though there was no history suggestive of heart failure, chronic liver disease (CLD), hypothyroidism, etc. He was recently evaluated for polyuria and polydipsia and was labeled as having uncontrolled DM with HTN. He gave no history of episodic palpitation, tremor, pallor, headache, sweating, easy bruising, striae, hyperpigmentation, back or joint pain, or any acute/chronic complications of DM.

Examination revealed a man with central obesity with thin limbs (BMI: 22.7 kg/m^2, waist circumference: 89 cm) having the classic lemon-on-stick appearance

S. Jahan, *Endocrinology Case Studies*,
https://doi.org/10.1007/978-981-95-2770-0_2

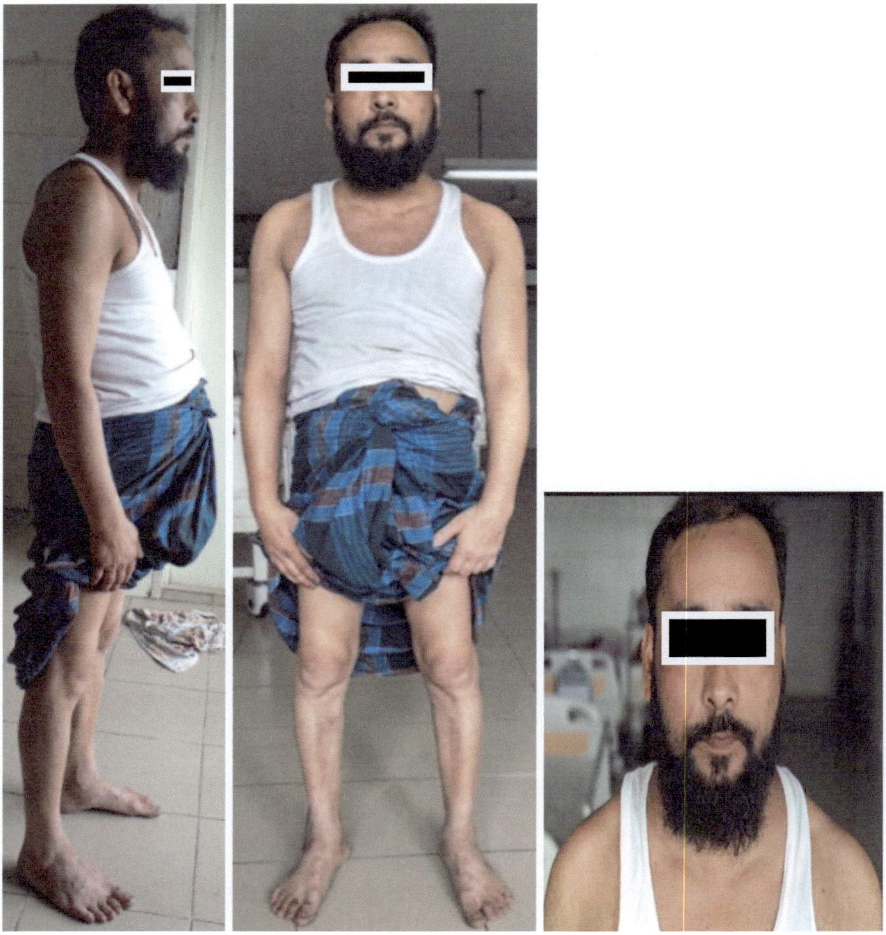

Fig. 2.1 Typical lemon-on-stick appearance of the patient

(Fig. 2.1). There was preferential fat deposition including fullness of supraclavicular fossa and dorsocervical fat pads. He had sinus tachycardia; blood pressure was 160/110 mm Hg without any postural drop. He also had thin skin, acanthosis nigricans in the neck, onychomycosis, tinea manuum, generalized tinea versicolor, palpable purpura, cracks in sole, and mild thyromegaly (Fig. 2.2).

Q1. What are the possible differential diagnoses in this man?
- (a) Metabolic syndrome
- (b) Cushing syndrome
- (c) Primary hypothyroidism
- (d) Chronic liver disease (CLD)
- (e) None of the above

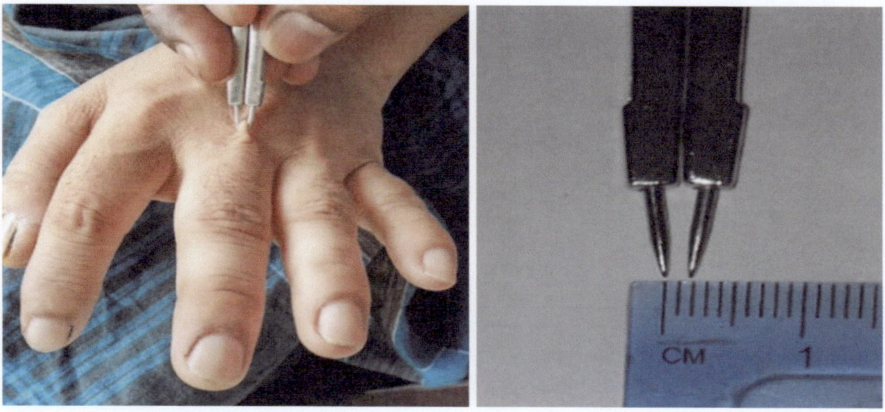

Fig. 2.2 Skin thinning measured by slide calipers

Table 2.1 Hematologic and biochemical profile of the patient

Parameter	Value	Reference range
CBC and ESR	Hb—**16.4 g/dL** Platelet—2,70,000/cmm TC of WBC: 13,000/cm Neutrophil: 80% Eosinophil: 0% Lymphocyte: 18% Basophil: 0%	Hb—15.2±2 gm/dL Platelet—150000–400000/cmm TC of WBC—7±3X10^9/l
HbA1C	6.2%	Normal ≤5.6%
S. Creatinine	0.72 mg/dL	0.5–1.3 mg/dL
SGPT	**102 U/L**	14–63 IU/L
Lipid profile	S. Cholesterol: 161, TG: 107, HDL-C: 38, DL-C: 101 mg/dL,	TC—250, TG—150, HDL—40, LDL—100 mg/dL
CPK	130 U/L	21–232 U/L
TSH	0.75 μIU/mL	0.35–5.50 μIU/mL

Q2. What specific features of Cushing syndrome were found in him?
- (a) Bruising and myopathy
- (b) Myopathy and striae
- (c) HTN and DM
- (d) Osteoporosis and hypokalemia
- (e) Osteoporosis and hypernatremia

The patient had a classic picture of Cushing syndrome and a negative history/clinical feature suggestive of other above differentials. He had some specific features of Cushing syndrome that included bruising, myopathy, HTN, osteoporosis, and hypokalemia. He was investigated accordingly to establish the diagnosis. His biochemical parameters were normal except having slightly elevated Hb with neutrophilic leucocytosis and fatty liver on abdominal USG (Table 2.1).

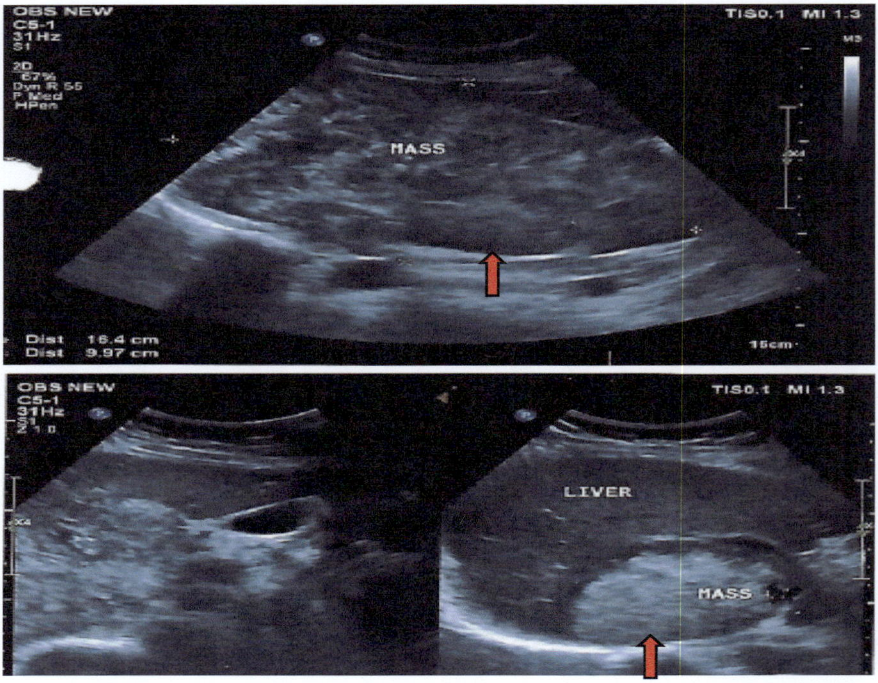

Fig. 2.3 Large, hypoechoic right adrenal mass on USG

Two months back, he had undergone abdominal imaging by an internist for evaluation of abdominal pain. USG of W/A showed a large (17 × 10) cm, hypoechoic right suprarenal mass for which he was referred to the endocrine unit (Fig. 2.3).

Q3. What are the possibilities for this large adrenal mass?
  (a) Adrenocortical carcinoma (ACC)
  (b) Malignant Pheochromocytoma
  (c) Adrenal metastasis
  (d) Myelolipoma
  (e) All the above

ACC was the probable diagnosis as the patient had rapidly progressive symptoms of relatively short duration with the feature of florid hypercortisolemia and a large adrenal mass. The absence of symptoms or signs related to the primary tumor (breast, bronchus, and renal) excluded the possibility of adrenal metastasis. Malignant pheochromocytoma would lead to HTN and local mass effect as in this case but would not lead to Cushing syndrome. As noted earlier, HTN was persistent, and hyperadrenergic spells were absent, making Pheo less likely. Myelolipoma would be clinically silent, only having a mass effect. So, ACC with hypercortisolemia was the clinical diagnosis.

Table 2.2 Hormonal profile of the patient showing hypercortisolemia and adrenal hyperandrogenemia

Test	Patient's value	Reference range	Cutoff value
ONLDDST (Cortisol)	**1082**	138–690 nmol/L	SCC >138 CS >50
24 h urinary free cortisol	**417.3**	50–190 μg/24 h	SCC >70 CS > 3xULN (>570)
Plasma ACTH	**<5**	0–46 pg/mL	<10: non-ACTH-dependent
24 h urinary metanephrine	**2080**	< 1775 nmol/day	>2X ULN (3550)
DHEAS	**754.1**	70–495 μg/ dL	

Q4. What investigations should be advised to establish the diagnosis of Cushing syndrome?
   (a) Serum electrolyte
   (b) Plasma ACTH
   (c) High-dose dexamethasone suppression test (HDDST)
   (d) Low-dose dexamethasone suppression test (LDDST)
   (e) 24 h urinary free cortisol (UFC)

Considering the high probability, the patient was screened for Cushing syndrome. The preferable screening tests include 24 h UFC, LDDST, midnight plasma cortisol, and late-night salivary cortisol. The 24 h UFC was > 2 times elevated, and on LDDST, the plasma cortisol remained nonsuppressed (Table 2.2). Other tests like serum electrolyte, HDDST, and plasma ACTH were done later to find out the etiology of Cushing syndrome once it was confirmed.

Q5. What other investigations are required to evaluate the adrenal mass?
   (a) Plasma aldosterone:renin activity ratio (ARR)
   (b) DHEA, DHEA-S
   (c) Plasma ACTH
   (d) 24 h urinary metanephrine
   (e) MIBG scan

The plasma ACTH level was suppressed, suggestive of adrenal etiology of Cushing; 24 h urinary metanephrine was mildly elevated possibly due to stress; DHEA was markedly high, indicating co-secretion of cortisol and adrenal androgen which is quite common in ACC (Table 2.2). PAC:PRA was not required here as the tumor in Conn's syndrome is usually 2 cm or less in size, and it is classically associated with hypokalemic HTN. MIBG is sometimes indicated in malignant pheochromocytoma to look for metastasis or possible paraganglioma that was not a possibility either.

Ultimately, the patient underwent a CT scan of the abdomen in the adrenal protocol that showed a large hypodense, regular mass (CC 10.4 cm, AP—12.5cm, TD—11.7cm, HU—40 U) with areas of calcification and necrosis, extending toward

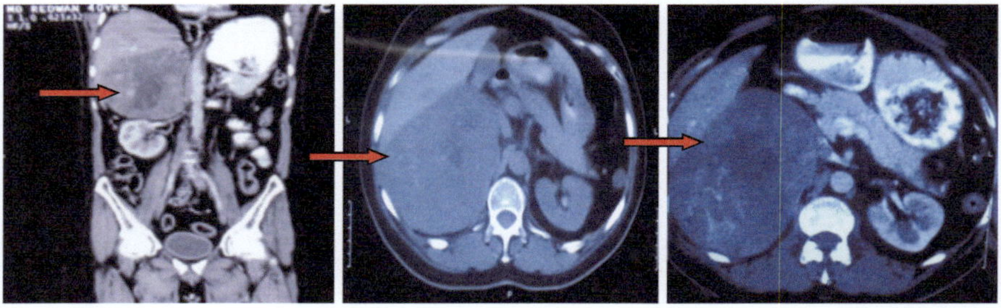

Fig. 2.4 Large, heterogeneous right adrenal lesion on CT abdomen

midline and compressing the inferior vena cava and superiorly the inferior surface of the right lobe of the liver (Fig. 2.4). After IV contrast, mild enhancement of the lesion and delayed washout (RWR <40%) were noted. The patient was later transferred into endocrine surgery and had a right adrenalectomy after adequate preparation. ACC was confirmed after the histopathologic examination of the resected tumor. He was later referred to the oncology department for further management.

Discussion

An adrenal lesion >4 cm in size, irregular, hypodense (HU>20 HU) in nature, having necrosis/calcification, and showing delayed washout on contrast administration is highly suggestive of ACC. The other two possibilities for such a lesion are adrenal metastasis, which is unlikely in the absence of primary malignancy elsewhere, and malignant pheochromocytoma, which is characterized by typical pheo spell and/or fluctuating HTN. Pheo cannot coexist with adrenal Cushing as their origin is diversely different (the exception is pheo with ectopic ACTH secretion where ACTH is expected to be high but not suppressed). The mean age of presentation of ACC at onset is between 40 and 50 years, although men tend to be older at presentation. Eighty percent of tumors are functional, most commonly secreting glucocorticoids alone (45%), glucocorticoids and androgens (45%), or androgens alone (10%). Fewer than 1% of all such tumors secrete aldosterone [1]. Patients present with features of the hormone excess state (glucocorticoid, androgen, or both), but abdominal pain, weight loss, anorexia, and fever occur in only 25% of cases. An abdominal mass may be palpable. Current treatment choices for what is often an aggressive tumor are poor. Surgery offers the only chance of cure for patients with local disease, but the metastatic spread is evident in 75% of cases at presentation. Radiotherapy is ineffective, as are most chemotherapeutic regimens. Mitotane in high doses offers benefit in reducing tumor growth and in controlling hormonal hypersecretion in 75% of cases. Overall, the prognosis is poor, with 5-year survival rates of less than 20% [2, 3].

Answers

Q1. (b)
Q2. (a), (d)
Q3. (a)
Q4. (d), (c)
Q5. (b), (c), (d)

References

1. Stewert PM, et al. The adrenal cortex. In: Melmed S, et al., editors. William's textbook of endocrinology. 12th ed. Philadelphia: Elsevier Saunders; 2011. p. 534–6.
2. Kasperlik-Zaluska AA, Migdalska BM, Zgliczynski S, et al. Adrenocortical carcinoma: a clinical study and treatment results of 52 patients. Cancer. 1995;75:2587–2591. 275.
3. Allolio B, Hahner S, Weismann D, et al. Management of adrenocortical carcinoma. Clin Endocrinol (Oxf). 2004;60:273–87.

3 A Young Lady with Progressive Neck Swelling and Bilateral Partial Ptosis

Introduction

A 24-year-old lady having the classic feature of hyperthyroid Graves' disease was noticed to have bilateral partial ptosis. She screened positive for Myasthenia Gravis on bedside ice on eyes test, and later the diagnosis was confirmed by positive autoimmune markers and by a repetitive nerve stimulation test. This case scenario is a unique example of clustering of endocrine and non-endocrine autoimmune diseases, and the subsequent discussion focuses on autoimmune polyendocrine syndrome.

Case Scenario

A 24-year-old lady presented with progressive neck swelling over 4 years that was associated with features of thyrotoxicosis. Six months later, she developed forward protrusion of both eyes along with redness, lacrimation, pain, and double vision. Two years back, her relatives noticed drooping of both of her eyelids. On inquiry, she admitted of marked diurnal variation of ptosis and restricted movement of the eyes in all directions. She also gave a history of occasional swallowing difficulty, particularly to liquid, and of nasal regurgitation. However, there was no history of limb weakness or breathing difficulty. None of her family members had a similar type of illness.

On examination, she was found to have warm, sweaty hands with fine tremor, her pulse rate was 120/min, regular, and BP was 120/50 mm of Hg. Neurological examination revealed bilateral partial ptosis (more on the right side) and reduced visual acuity (6/9 on both eyes), complete external ophthalmoplegia, intact light and accommodation reflex, and a normal fundoscopic examination. Her deep tendon reflexes were exaggerated, though the plantar response was flexor. She had a diffuse, firm, smooth-surfaced, nontender thyromegaly with a bruit but without any retrosternal extension (Fig. 3.1).

S. Jahan, *Endocrinology Case Studies*,
https://doi.org/10.1007/978-981-95-2770-0_3

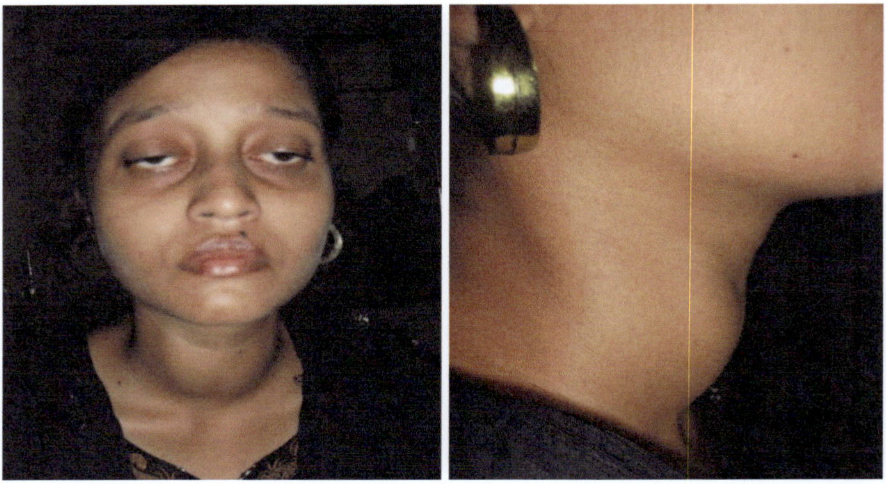

Fig. 3.1 Lady with bilateral partial ptosis and diffuse goiter. (Photograph was taken with prior consent from the patient)

Q1. What are the probable clinical diagnoses for this lady?
   (a) Hyperthyroid Graves' disease
   (b) Autoimmune polyendocrinopathy type-1 (APS type-1)
   (c) Myasthenia Gravis (MG)
   (d) Autoimmune polyendocrinopathy type-2 (APS type-2)
   (e) Ocular myopathy

The clinical picture was compatible with APS type-2. Hyperthyroid Graves' disease on itself would not be responsible for the partial ptosis. Likewise, only having Myasthenia Gravis was insufficient to explain diffuse thyromegaly with toxicity and proptosis. She had a combination of both the conditions as in APS type-2. APS type-1 usually occurs much earlier (between 5 and 15 years) and is characterized by a triad of mucocutaneous candidiasis, primary hypoparathyroidism, and Addison's disease, often in combination with ectodermal dysplasia. Ocular myopathy is likely to manifest at an earlier age in a person with a positive family history, and it is unlikely to be associated with any diurnal variation or double vision as in this case.

Q2. What convenient bedside test can be done to establish the diagnosis?
   (a) Tensilon test
   (b) Ice on eyes test
   (c) Pemberton's test
   (d) Valsalva maneuver
   (e) None of the above

Ice on eyes test is a very convenient bedside test with good sensitivity and specificity for screening Myasthenia Gravis where ice pack is kept on the eyelids for

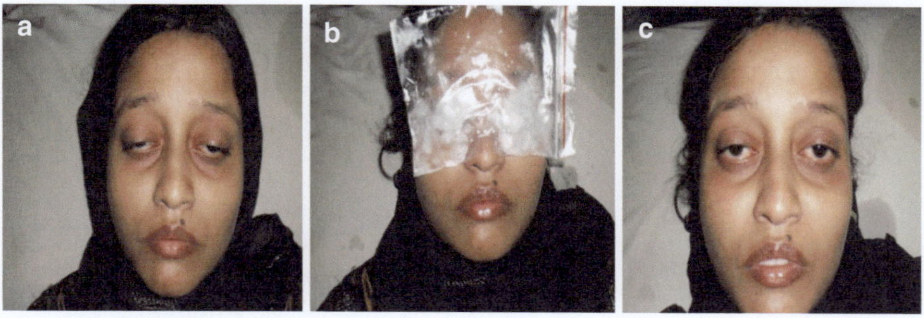

Fig 3.2 Ice on eyes test (panel A—palpebral fissure, 5 mm, panel B—ice on eyes, panel C—palpebral fissure, 12 mm)

2 min and is deemed positive if there is an improvement of ptosis or increase of palpebral fissure by 2 mm (rule of two). Tensilon test is cumbersome and hazardous So, it is rarely performed nowadays for the screening of MG. On the other hand, Pemberton's test is done to exclude retrosternal extension of a large goiter that is rather unnecessary in this case. Valsalva maneuver is carried out to look for cardiac autonomic neuropathy and is not indicated in our case. She underwent the ice on eyes test with a remarkable improvement of her ptosis (Fig. 3.2).

Q3. What abnormalities might be expected on routine investigations of this patient?
   (a) Raised muscle enzyme
   (b) Raised liver enzymes
   (c) Hyponatremia and hyperkalemia
   (d) Renal glycosuria
   (e) Lag storage glycosuria

Thyrotoxicosis is associated with raised liver enzymes (SGPT/SGOT/ALP/γ-GT), whereas hypothyroidism and myopathies are associated with raised muscle enzymes. Hyponatremia occurs in hypothyroidism, and hyperkalemia classically happens in Addison's disease. Thyrotoxicosis leads to rapid glucose absorption from the gut, causing an early peak of blood glucose spilling into urine that levels off quickly; the term applied is lag storage glycosuria. Renal glycosuria typically occurs in pregnancy where the renal tubular threshold for glucose reabsorption is lowered.

She was found to have suppressed TSH with very high FT4 and FT3. Thyroid antibodies including anti-TPO, anti-TG, and TRAb were positive in high titer (Table 3.1). Thyroid scan revealed a markedly enlarged thyroid with a uniform radiotracer concentration and absent background activity (Fig. 3.3). Though her chest X-ray failed to show any abnormality, HRCT of the chest revealed enlarged thymus subsequently. On repetitive nerve stimulation, there was >15% decremental response with post-exercise facilitation (in left orbicularis oculi). Anti-acetylcholine receptor Ab (ACRA) was also found to be positive in high titer (Table 3.1).

Table 3.1 Biochemical report of the patient's severe toxicosis with autoimmunity

Parameter	Patient's value	Reference range
TSH	**<0.01 μIU/mL**	0.4–4
FT4	**60.64 pmol/L**	9.13–23.8
FT3	**30.0 pgm/mL**	1.8–4.7
Anti -TPO Ab[a]	**637.78 IU/mL**	Up to 35
Anti-TG Ab[b]	**>1000 IU/mL**	Up to 40
TRAb[c]	**15 miu/mL**	1.5 miu/mL
ACRA[d]	**15.5 nmol/L**	<0.4 nmol/L

[a]Anti -TPO Ab—Antithyroperoxidase antibody
[b]Anti TG Ab—Antithyroglobulin antibody
[c]TRAb—Thyroid receptor antibody
[d]ACRA—Anti-acetylcholine receptor antibody

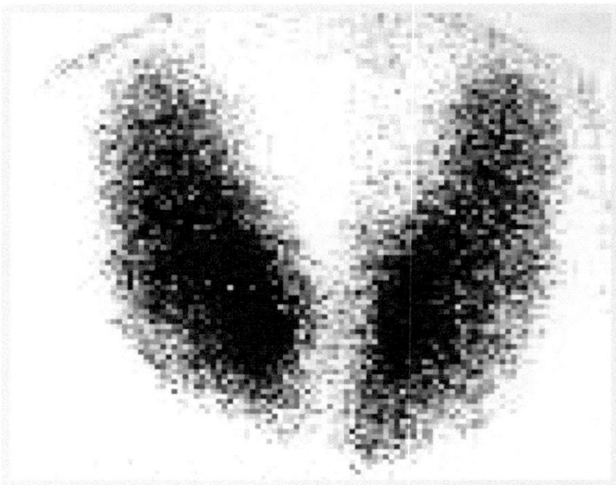

Fig. 3.3 Thyroid scan showing a diffuse, high uptake of a radio tracer in the thyroid bed

Q4. What other investigations are required to exclude the associated autoimmune diseases?
(a) Serum prolactin
(b) IGF-1
(c) Basal cortisol, plasma ACTH
(d) Growth hormone stimulation test
(e) Short Synacthen test

APS type-2 is characterized by autoimmune thyroiditis, Addison's disease, type-1 DM, premature gonadal failure, and several other nonendocrine autoimmune diseases. Thereby, basal cortisol, plasma ACTH, as well as a short Synacthen test are required to exclude adrenal insufficiency. Other investigations stated above are unnecessary in this case. They are usually done in the setting of pituitary tumors like prolactinoma and GH deficiency or excess (acromegaly). Though she screened negative for other autoimmune diseases, still she needs to be monitored on a regular basis lifelong.

The patient was started on the tab. Carbimazole 45 mg and tab. pyridostigmine 60 mg daily. After being euthyroid, she underwent thymectomy. Currently, she is on regular follow-up and doing well.

Discussion

The prevalence of *autoimmune thyroid disease (ATD)* is high and estimated to be 5% in the general population [1], while *Myasthenia Gravis (MG)* is a rare disease with an incidence of 8–10 cases per million persons/year and a prevalence of 150–250 cases per million [2, 3]. ATD is one of the most representative organ-specific autoimmune disorders; it is associated with other autoimmune endocrine failures or nonendocrine diseases [4]. The clustering of such prototype immune-mediated disorder leads to various polyendocrinopathies like APS type-1 or APS type-2. Among the nonendocrine diseases' vitiligo, pernicious anemia, MG, autoimmune gastritis, celiac disease, and hepatitis are noteworthy [5, 6]. The most common diseases coexisting with MG are GD and HT, with a frequency of 7 and 3%, respectively [7]. When associated with ATDs, MG shows specific features, such as the young age of onset, mild clinical symptoms, low levels of ACR antibodies, and low frequency of thymic alterations. Although they occur concurrently due to the same underlying pathophysiologic defect, their treatment is quite different. Myasthenia is treated with anticholinesterase drugs and immunosuppressive agents like azathioprine, steroids, and monoclonal antibodies against B lymphocytes, whereas GD is treated with antithyroid drugs or radioiodine ablation. Surgery is applicable in both cases, including thymectomy in associated thymoma/thymic hyperplasia in MG and thyroidectomy in GD.

Answers

Q1. (d)
Q2. (b)
Q3. (b), (c)
Q4. (c), (e)

References

1. Hollowell JG, Staehling NW, Flanders WD, Hannon WH, Gunter EW, Spencer CA, et al. Serum TSH, T (4), and thyroid antibodies in the United States population (1988 to 1994): National Health and Nutrition Examination Survey (NHANES III). J Clin Endocrinol Metab. 2002;87:489–99.
2. Gilhus NE. Myasthenia gravis. N Engl J Med. 2016;375:2570–81. https://doi.org/10.1056/NEJMra1602678.

3. Antonelli A, Ferrari SM, Corrado A, Di Domenicantonio A, Fallahi P. Autoimmune thyroid disorders. Autoimmun Rev. 2015;14:174–80.
4. Betterle C, Dal Pra C, Mantero F, Zanchetta R. Autoimmune adrenal insufficiency and autoimmune polyendocrine syndromes: autoantibodies, autoantigens, and their applicability in diagnosis and disease prediction. Endocr Rev. 2002;23:327–64.
5. Weetman AP. Diseases associated with thyroid autoimmunity: explanations for the expanding spectrum. Clin Endocrinol (Oxf). 2011;74:411–8.
6. Cellini M, Santaguida MG, Virili C, Capriello S, Brusca N, Gargano L, et al. Hashimoto's thyroiditis and autoimmune gastritis. Front Endocrinol. 2017;8:92.
7. Mao ZF, Yang LX, Mo XA, Qin C, Lai YR, He NY, et al. Frequency of autoimmune diseases in myasthenia gravis: a systematic review. Int J Neurosci. 2011;121:121–9.

4 A 38-Year-Old Farmer Presented with Epigastric Pain and Drowsiness

Introduction

A 38-year-old farmer having a brief history of weight loss and epigastric pain developed severe hypercalcemia culminating in an acute crisis necessitating urgent hospitalization. He was disoriented, severely dehydrated, and had a tender hepatomegaly. Investigations revealed hepatocellular carcinoma (HCC) on the top of cirrhosis of liver, severe hypercalcemia with suppressed PTH, and pre-renal azotemia. Finally, he was diagnosed with paraneoplastic hypercalcemia related to PTH-related peptide secretion from HCC. This is a unique case of paraneoplastic hypercalcemia arising out of epithelial carcinoma, and the review literature highlights the feature of hypercalcemia of malignancy.

Case Scenario

A 38-year-old farmer presented with mild upper abdominal pain for 2 months. It was of gradual onset, constant, dull aching in nature, aggravated by food and partially relieved by an anti-ulcerant. The pain was nonradiating in nature and not associated with any abdominal distention. He had two to three episodes of vomiting over the last 1 month. The vomitus was small in amount, contained food materials, not blood or bile-stained, and was nonprojectile in nature. He had an unintentional weight loss of about 10 kg during this period. For the last 15 days, he gradually developed an altered level of consciousness. It was preceded by irrelevant speech and sleep disturbance. Finally, at the time of admission, he was found to be drowsy and disoriented. There was no H/O melena, hematochezia, hematemesis, jaundice, ascites, loss of libido, blood transfusion, surgical intervention, needle prick, or illicit sexual exposure. There was also no H/O fever, cough, shortness of breath, chest pain, polyuria, polydipsia, or loin pain. He had H/O smoking of at least 10 packs a year but no H/O alcohol or any substance abuse. He underwent investigations to

S. Jahan, *Endocrinology Case Studies*,
https://doi.org/10.1007/978-981-95-2770-0_4

Table 4.1 Sequential hematologic reports of the patient showing anemia with neutrophilia

Date	20-09-17	14-10-17	24-10-17	25-10-17
Hb %	12.8	13.7	14.1	13.5
ESR	60	**76**	**90**	25
TC	12,500	**21000**	**21,500**	**20,000**
RBC	–	5.7	6.06	5.66
Platelets	–	300,000	210,000	200,000
Neutrophils	**75**	**83**	**85**	**84**
Lymphocyte	17	10	08	10
Monocytes	03	02	06	04
Eosinophil	05	05	01	02
Basophil	0	00	00	00
PBF	–			Neutrophilic leukocytosis

Table 4.2 Biochemical report of the patient showing severe hypercalcemia with suppressed PTH

Date	24/10/17	25/10/17	30/10/17	31/10/17	1/11/17	04/11/17
Calcium (mg/dl)	**19.8**	**20.7 Calculated**	**15.2**	**14.8**	**12**	**11.9 Calculated**
Albumin (gm/L)	-	35	-			26
PO4(mg/dL)	4.9	3.7	4.5			
PTH(pg/mL)		**7.7**	6.6			
ALP(U/L)		368				
Serum 1,25(OH)$_2$vit-D						89 pg/ml
S.creatinine (mg/dL)	2.35	3.34	2.18	2.14	1.92	1.56
Serum uric acid						11.5 mg/dl
Serum electrolyte	Na—135, K—3.6, Cl—96, TCO_2—27 mmol/l					
Urine R/E	RBC—nil Pus cells—plenty					
eGFR (ml/min/1.73 m2)				36.6	37	42

evaluate these complaints. The hematologic report showed neutrophilic leukocytosis with elevated ESR (Table 4.1). The biochemical report showed severe hypercalcemia with raised serum creatinine (Table 4.2). Then, he was referred to the endocrine unit for the evaluation of hypercalcemia.

On examination at the endocrine unit, he was found to be drowsy (GCS-12/15), his BMI was 20.80 kg/m^2, pulse rate—70/min, and BP—100/70 mm Hg. He was severely dehydrated, mildly anemic, non-icteric, and non-edematous. There were no stigmata of CLD or CKD except moderate leukonychia. The flapping tremor was absent. The abdomen was distended with fullness of the flanks, and umbilicus was centrally placed and inverted with a transverse slit. A tender hepatomegaly of 10 cm from the xiphoid process was noted. It was hard in consistency, smooth-surfaced with a regular margin, not associated with a bruit. Ascites were absent.

Q1. What was the probable reason behind the impaired level consciousness in this patient?
  (a) Hepatic encephalopathy
  (b) Uremic encephalopathy
  (c) Electrolyte imbalance
  (d) Severe hypercalcemia
  (e) None of the above

Provisional diagnosis at this point was hypercalcemic crisis due to primary hyperparathyroidism or hypercalcemia of malignancy. However, the tender hepatomegaly remained unexplained at this stage, and the patient did not give any H/O jaundice, hematemesis, ascites, or any source of hepatic viral acquisition like blood transfusion, operations, and I/V drug abuse linking it to CLD. Physical examination failed to reveal any stigmata of CLD except tender hepatomegaly or signs of acute hepatic decompensation (flapping tremor). On the other hand, features suggestive of CKD, e.g., anorexia, oliguria, nocturia, edema, pruritus, as well as contributory etiologies (DM, HTN, and vasculitis) to it, were also absent. He had only a few episodes of minimal vomiting that were unlikely to account for his symptoms, and it was further supported by a normal electrolyte report (Table 4.2). Hypercalcemia is arbitrarily divided into mild (<12 mg/dL), moderate (12–14 mg/dL), and severe categories (>14 mg/dL). Our patient had severe hypercalcemia culminating in hypercalcemic crisis necessitating urgent admission. At this point, he was aggressively treated with I/V N/S (4–6 L/day), inj. hydrocortisone along with I/V broad-spectrum antibiotics. Once dehydration was corrected, the diuresis was enforced with inj. furosemide 20 mg b.d. However, I/V bisphosphonate could not be given because of impaired renal function (eGFR- <36.6 mL/min). Expectant management of hepatic encephalopathy was also given as it could not be excluded entirely on clinical ground.

Following resuscitation, he was extensively investigated and was found to have neutrophilic leukocytosis with raised ESR on several occasions (Table 4.1). He also had mild hyperuricemia and evidence of UTI on urine R/E and C/S. The change in his renal function and calcium panel occurring during hospital course is shown in Table 4.2.

Q2. What is the likely etiology of hypercalcemia in this patient?
  (a) Primary hyperparathyroidism
  (b) Hypercalcemia of malignancy
  (c) Secondary hyperparathyroidism
  (d) Tertiary hyperparathyroidism
  (e) Tumor lysis syndrome

Primary hyperparathyroidism was unlikely as the presentation was dramatic with the acute onset of severe hypercalcemia. Primary hyperparathyroidism tends to present in an asymptomatic or insidious manner typically in a middle-aged female with mild-to-moderate hypercalcemia. This is the commonest cause of hypercalcemia presenting in endocrine OPD. Elevated PTH with high serum calcium and low phosphate is the classic biochemical finding in such a case. Secondary hyperparathyroidism occurs in the context of Vit-D deficiency, CKD, and presents with hypocalcemia rather than hypercalcemia. Hypercalcemia of tertiary hyperparathyroidism presents at an advanced stage in CKD with florid symptoms along with high phosphate levels. Tumor lysis syndrome occurs in some hematologic or solid malignancies and tends to present abruptly with acute renal shutdown with severe hyperphosphatemia, high CPK, and raised serum creatinine but with low serum calcium. Finally, hypercalcemia of malignancy was the likeliest diagnosis as it necessitates emergency hospital admission and resuscitation. So, he had hypercalcemia of malignancy due to either primaries/secondaries in the liver that led to a very dramatic presentation of severe hypercalcemia and was associated with suppressed PTH. So, the next investigations were planned accordingly.

Q3. What should be the next evaluation plan for him?
(a) USG of whole abdomen
(b) Skeletal survey
(c) PTH-related peptide assay
(d) 25(OH) Vit-D
(e) CXR P/A view

USG of W/A revealed hepatomegaly with coarse echotexture and the presence of multiple SOL (suggestive of multifocal hepatocellular carcinoma (HCC)/metastasis), bilateral renal parenchymal disease, huge ascites, and bilateral pleural effusion (left>right). Skeletal surveys as well as CXR P/A were unremarkable. 1, 25-dihydroxy Vit-D level was found to be elevated. Plasma protein electrophoresis revealed polyclonal gammopathy. PTH-related peptide assay could not be done due to the lack of local availability. 25(OH) Vit-D reflects its total storage form and is assayed in cases of vitamin D deficiency disorder. Other investigations established the diagnosis of HCC on the top of cirrhosis, whereby hypercalcemia was probably paraneoplastic (Table 4.3).

Table 4.3 Investigations establishing HCC on CLD (HBV origin)

Investigations	Patient's report	Reference value
AFP	**159.24 ng/mL**	<1 ng/mL
Serum CEA	2.18 ng/mL	Smoker up to 8 ng/mL
HBsAg	Positive	–
anti- HCV Ab	Negative	–
PT	14 s	
Upper GI endoscopy	Congestive gastropathy with erosive antral gastritis	

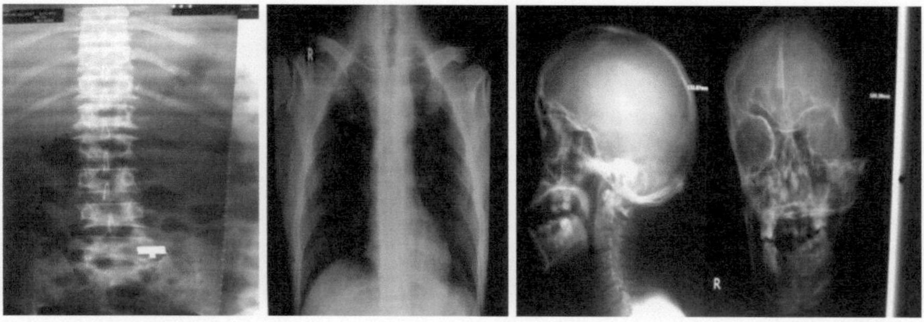

Fig. 4.1 Unremarkable skeletal survey of the patient

So, our final diagnosis was CLD with HCC with paraneoplastic hypercalcemia possibly related to PTH-related peptide and pre-renal AKI. Ultimately, the patient was referred to hepatic oncology, and he succumbed to death a few months later (Fig. 4.1).

Discussion

Hypercalcemia is usually a terminal event in malignancy. At the time of presentation, the patient generally has a full-blown picture of primary malignancy, and hypercalcemia is an added complication. Squamous cell carcinoma of the lungs is a prime example. The prototype patient is a middle-aged smoker presenting with new-onset cough, chest pain, SOB, hemoptysis, weight loss, or with the feature of distant metastasis like lymphadenopathy, hepatosplenomegaly, bone pain, jaundice, etc. Patients present with the classic signs and symptoms of severe hypercalcemia or hypercalcemic crisis like confusion, polydipsia, polyuria, constipation, nausea, and vomiting. Perhaps because of the acuteness of hypercalcemia and the elderly patient population involved, dramatic changes in mental status, culminating in a coma, are relatively common. The diagnosis can be missed because the manifestations often overlap those of the underlying malignancy and because low serum albumin may lead to a normal serum total calcium value despite an elevated blood-ionized calcium concentration [1]. Malignancy is implicated in hypercalcemia in three ways. The commonest is the squamous cell carcinoma of the head and neck and multiple myeloma with extensive skeletal metastasis causing bone resorption leading to high serum calcium and phosphate with suppressed PTH. The second category of tumors like lymphoma has 1-α-hydroxylase activity within the neoplastic cell causing excess 1, 25-dihydroxy Vit-D activity and high serum calcium and phosphate with suppressed PTH. Such enzymatic activity is also noticed within a multinucleated giant cell in granulomatous disease like sarcoidosis. On the other hand, secretion of PTH-related peptides by a tumor as in squamous cell carcinoma of lungs leads to high serum calcium and low/low normal phosphate and suppressed PTH. This can be diagnosed by a specific assay for the detection of PTH-related peptides. As this tumor product acts similar to endogenous PTH, it will lead to increased conversion of 25(OH) Vit-D into active metabolite 1, 25-dihydroxy-Vit-D, and it is used as a supportive marker of such tumor. Epithelial carcinomas like renal cell carcinoma, hepatocellular carcinoma, ovarian cancer, and colon cancer may present with hypercalcemia as a paraneoplastic syndrome due to the elaboration of PTH-related peptides by them. As our patient had hypercalcemia with low phosphate, elevated 1, 25-dihydroxy Vit-D, suppressed PTH, and the absence of bone involvement/metastasis, the possible diagnosis was hypercalcemia due to PTH-related peptide secretion. In rare instances, the tumor itself can lead to ectopic production of PTH which simulates the biochemical picture of primary hyperparathyroidism. Once the patient develops hypercalcemia, the prognosis is dismal, and only 10–20% survive beyond 1–2 years [1].

Answers

Q1. (d)
Q2. (b)
Q3. (a), (b), (c), (e)

Reference

1. Bringhurst FR, et al. Hormones and disorders of mineral metabolism. In: Melmed S, et al., editors. William's textbook of endocrinology. 12th ed. Philadelphia: Elsevier Saunders; 2011. p. 1270–1.

5 A 60-Year-Old Farmer Presented with Anorexia, Weight Loss and Bilateral Adrenal Incidentaloma

Introduction

The discovery of bilateral adrenal incidentaloma in an elderly male having nonspecific symptoms of weight loss and abdominal pain raises the suspicion of adrenal tuberculosis, histoplasmosis, or other deep fungal infection, malignancy, or metastasis. In an endemic region of TB, disseminated histoplasmosis is often misdiagnosed and treated as TB. So, a high index of suspicion is mandatory while dealing with such a case where the ultimate diagnosis relies on adrenal FNAC and histopathology. The following scenario highlights such a unique presentation of disseminated histoplasmosis where the initial inconclusive histopathology created much confusion before the final diagnosis could be made by reviewing the reports.

Case Summary

A 60-year-old farmer presented with mild-to-moderate abdominal discomfort for 3 months. He experienced occasional dull aching, nonradiating pain in the upper abdomen that was not associated with food, and was partially relieved by an anti-ulcerant. The pain was associated with slight abdominal fullness. He complained of anorexia and nausea during this period and noticed unintentional weight loss that was not documented. However, it was evidenced by the loosening of his clothes. There was no history of melena, hematochezia, hematemesis, jaundice, ascites, altered consciousness, loss of libido, etc. He did not complain of fever, night sweats, cough, shortness of breath, chest pain, contact with sputum-positive TB patient, or any bowel or bladder abnormality. He was investigated at a tertiary care hospital and found to have anemia with high ESR, hypoalbuminemia and an altered A:G ratio, hyponatremia, and an ill-defined patchy opacity at the right apical region on CXR (Table 5.1 and Fig. 5.1).

S. Jahan, *Endocrinology Case Studies*,
https://doi.org/10.1007/978-981-95-2770-0_5

Table 5.1 Biochemical profile of the patient showing anemia with high ESR

Parameter	14-09-17	25-09-17	10.11.17
Hb %	12.4	12.7	**12.8**
ESR	**110**	**136**	**35**
TC	6200	6000	5000
Platelets	2,20,000	2,30,000	1,80000**16,810,0**
Neutrophils	58	59	55
Lymphocytes	38	34	35
Monocytes	02	05	07
Eosinophil	02	02	03
Basophil	0	00	0
PBF	Mild hypochromic anemia		**1,801,00011**
Creatinine (mg/dL)	0.9		
HbsAg	Negative		
Total protein (g/L)		83	
ALT (U/L)		31.6	
Albumin (g/L)		24	
Globulin(g/L)		59	
Albumin:Globulin (A: G) ratio		**0.4:1**	
AFB		Negative	
Gene Xpert		MTB not detected	
Chest X-ray P/A view	NAD	**Ill-defined opacities in the right upper zone suggestive of inflammation**	

Q1. What should be the next plan for investigations?
 (a) Myeloma screen
 (b) Tuberculosis screen
 (c) USG of W/A
 (d) CT-guided FNAC of lung lesion
 (e) UGIT endoscopy

As this elderly patient presented with weight loss, anorexia, very high ESR, and an altered A:G ratio, a myeloma screen was mandatory. However, he did not have fever, bone pain, or body ache, and investigations failed to reveal any rouleaux formation/atypical plasma cells, skeletal lesion, or altered renal function, making the diagnosis unlikely. Plasma protein electrophoresis showed polyclonal gammopathy. Considering his Asian ethnicity, mode of presentation, and very high ESR with pulmonary opacity, TB was the likeliest provisional diagnosis though the serologic tests were found to be normal (sputum for AFB—negative; Gene X-pert—MTB not detected; MT—negative). Elderly male with abdominal pain, anorexia, and asthenia certainly warrants search for occult GIT malignancy. UGIT endoscopy only revealed

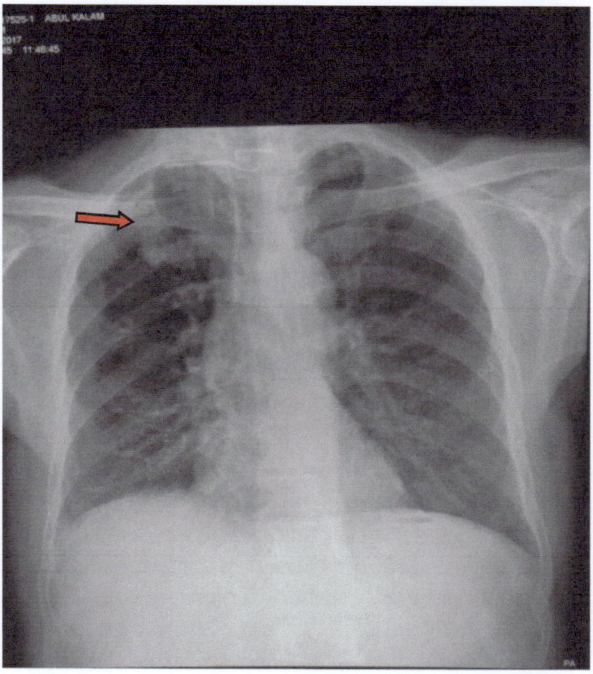

Fig. 5.1 Ill-defined patchy opacity of right apical region on CXR

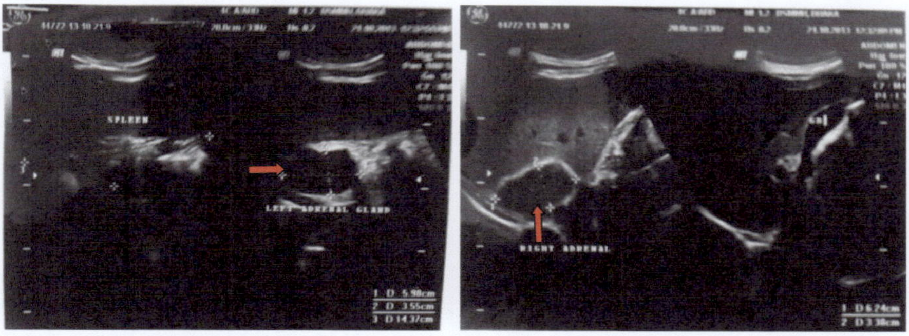

Fig. 5.2 USG of W/A showing bilateral hypoechoic suprarenal masses

pre-pyloric lesion. CT-guided FNAC of the pulmonary lesion would be in the next tier of investigations if others failed to detect the diagnosis. Meanwhile, there was an interesting finding on USG of W/A that showed bilateral adrenal incidentaloma. These were hypoechoic, uniform soft tissue masses measuring (8 × 3.6) and (6 × 2.8) cm on the right and left sides, respectively (Fig. 5.2). They were associated with mild hepatosplenomegaly and a small echogenic right kidney without any lymphadenopathy or ascites. Now at this point, the diagnosis took an abrupt turn into bilateral adrenal incidentaloma with a pulmonary lesion.

Q2. What are the differential diagnoses at this point?
(a) Cushing syndrome
(b) Conn syndrome
(c) Pheochromocytoma
(d) Nonfunctioning adrenal adenoma
(e) Adrenal metastasis

All the above differentials were remote possibilities because the patient did not have any stigmata of Cushing syndrome (bruises, myopathy, HTN, striae, central obesity, and osteoporosis), and adrenal tumor in Cushing syndrome tends to be a small, unilateral adenoma. However, subclinical Cushing syndrome at times may present without any telltale sign of the disorder, but it commonly manifests with metabolic abnormalities like DM, HTN, or osteoporosis that were absent in this case. On the other hand, the absence of HTN and hypokalemia made the possibility of Conn syndrome unlikely where the adrenal nodule is commonly unilateral and < 2 cm in size. Malignant, bilateral pheochromocytoma could have been possible, provided there were other clues like paroxysmal/persistent HTN, hyperadrenergic spell, H/O pressor response during anesthesia, etc. None of the above entities were likely to explain the chest and other clinical/biochemical abnormalities. So, nonfunctioning adenoma of this size also seemed unlikely. Primary tumor in the lung with bilateral adrenal metastasis was one of the likely diagnoses, and investigations were arranged accordingly.

Q3. What should be the next plan of investigations?
(a) 24 h urinary free cortisol
(b) 24 h urinary metanephrine
(c) Basal cortisol/short synacthen test
(d) CT scan of the adrenals/FNAC of the adrenal mass
(e) FNAC of the lung lesion

The hormonal reports including basal and stimulated cortisol before admission into our unit were normal. The reports of the adrenal CT scan and FNAC were also available at the time of admission. The adrenal CT scan showed bilateral large, well-defined, heterogeneously enhanced suprarenal masses [right—(5.6 × 3.6 × 6.5) cm and left—(4.7 × 3.5 × 6.3 cm)] with a central area of necrosis and calcification (Fig. 5.3). The histopathology slide revealed several cuboidal and uniform-looking epithelial cells arranged in cohesive clusters that contained round nuclei with scanty cytoplasm. The features were suggestive of an adrenal adenoma, possibly Cushing syndrome.

At this point, when exhaustive investigations failed to reach a definite diagnosis, he was referred to our department. We doubted the previous histopathology report as functioning or nonfunctioning adrenal adenoma was a rare possibility as already discussed. So, before embarking on further investigations, differentials needed to be reconsidered rationally.

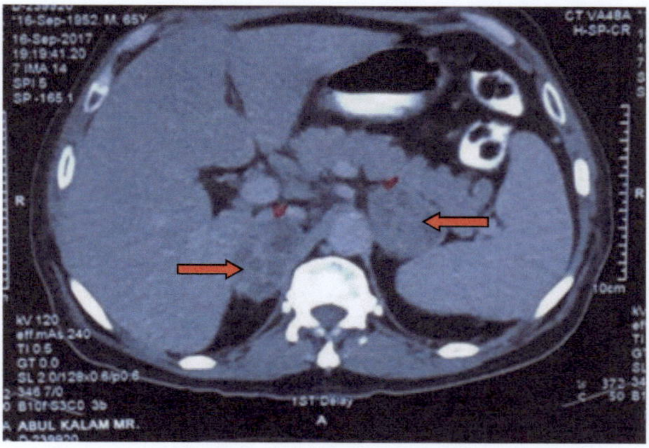

Fig. 5.3 Bilateral, large heterogeneous adrenal masses

Table 5.2 Hormonal parameter of the patient showing high ACTH and normal cortisol

Parameter	2.10.17	12.11.17
Basal cortisol	461	313
Stimulated cortisol	452.17 nmol/L (at 30 min) 479 nmol/L (at 60 min)	380 (at 30 min) 420 (at 60 min)
Plasma ACTH	96 pg/mL	101 pg/mL
DHEA-S		25
24 h urinary metanephrine		1067

Q4. What are the possibilities now?
- (a) Adrenocortical carcinoma
- (b) Adrenal metastasis
- (c) Adrenal tuberculosis
- (d) Adrenal fungal infection (histoplasmosis)
- (e) Infiltrative adrenal diseases

We were dealing with large, bilateral, irregular adrenal masses having areas of necrosis and calcification (density on noncontrast CT-40 HU). So, differentials included malignancy, either primary or secondary, and infective/infiltrative lesion. Adrenocortical carcinoma commonly presents with features of hormonal hypersecretion, e.g., hypercortisolism and hyperandrogenism, besides the usual feature of malignancy. The patient underwent hormonal evaluation accordingly. After admission, he had several episodes of hematemesis and was transfused with 1 unit red cell concentrate (RCC). This time, he had high ACTH with subnormal cortisol response on stimulation, indicating functional adrenal insufficiency in the context of an acute episode of bleeding. His DHEA-S was also subnormal, indicative of adrenal insufficiency (Table 5.2). These tests are mandatory in case of bilateral adrenal lesions that might compromise adrenal function if > 90% of the gland is destroyed. 24 h urinary metanephrine is required before performing FNAC or other procedures to effectively rule out pheochromocytoma, thereby averting life-threatening

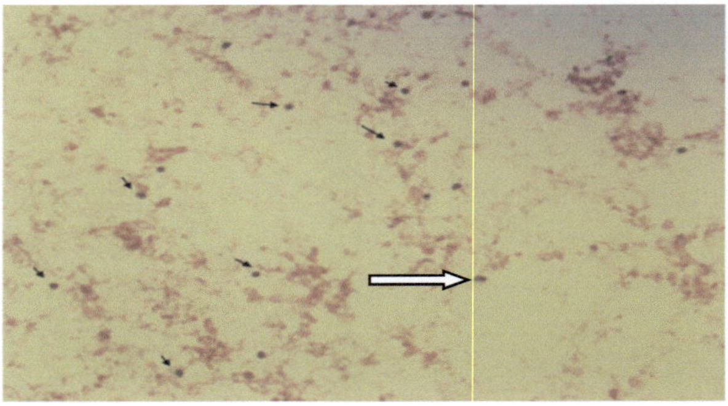

Fig. 5.4 Histopathology of the adrenal tissue showing budding yeast of *H. capsulatum*

hypertensive crisis. However, 24 h UFC is not the choice of a test as it is done only to screen cortisol excess, and in cases of suspected insufficiency as in this case, the filtered free fraction would be too low to be detected in urine. Considering the fact that the patient was a nonsmoker having no H/O cough, chest pain, and hemoptysis and that in only 10–20% of cases the adrenal metastases are bilateral, most of them being <4 cm, it also seemed less likely. At the same time, CXR revealed patchy opacity instead of showing a solitary nodule that was in keeping with TB/fungal infection.

Finally, we decided to review the histopathology slide and performed CT-guided adrenal FNAC at our center. This time, the report confirmed our suspicion, with slides showing mostly coagulative necrosis, cluster and single histiocytes, and few inflammatory cells. Several macrophages showed *Histoplasma capsulatum* within their cytoplasm. There was no evidence of any adrenal adenoma or malignant cell (Fig. 5.4). So, the final diagnosis was disseminated histoplasmosis. The patient was started on tab. Itraconazole 400 mg daily, continued for 2 years, with the complete remission of clinical and laboratory abnormalities.

Discussion

Disseminated histoplasmosis is a rare and potentially fatal disease caused by dimorphic soil fungus *Histoplasma capsulatum* [1]. During dissemination, it frequently (80%) affects the adrenal glands [2]. In such cases, patients often present with bilateral adrenal masses [3]. Adrenal histoplasmosis must be followed particularly carefully due to the possibility of advancing to the point of causing adrenal insufficiency—the most common cause of death in patients with disseminated Histoplasma infection [4, 5]. The clinical and radiologic features are like those seen in tuberculosis (TB), malignancy or adrenal metastasis, and other deep fungal infections. In the endemic areas of TB, it is frequently misdiagnosed and treated with antitubercular drugs. The only investigation that is confirmatory in confusing cases is a demonstration of fungal components in adrenal tissue derived from ultrasonography/computed tomography (USG/CT)-guided fine needle aspiration cytology (FNAC). Immunosuppression, steroid therapy, hematologic malignancies, advanced age, and chronic medical diseases predispose to disseminated histoplasmosis [6]. The cause of Histoplasma tropism for the adrenal gland is unclear, but it is hypothesized that the glucocorticoid-rich adrenal cells and the relative scarcity of reticuloendothelial cells may contribute [6]. Adrenal histoplasmosis should be treated like disseminated histoplasmosis. The mortality in untreated patients is as high as 80–100%, but if treated with antifungal drugs, this very high mortality rate is remarkably reduced to less than 25% [7]. Patients who are not severely ill can be treated with oral itraconazole, 200 mg bid. Patients with severe infection should be treated initially with amphotericin B at a dosage of 0.7 to 1 mg/kg daily or the lipid formulation of amphotericin B at a dosage of 3 to 5 mg/kg daily [5].

Answers

Q1. (a), (b), (c), (d), (e)
Q2. (c)
Q3. (b), (c), (d)
Q4. all.

References

1. Rana C, Krishnani N, Kumari N. Bilateral adrenal histoplasmosis in immunocompetent patients. Diagn Cytopathol. 2011;39(4):294–6. https://doi.org/10.1002/dc.21416.
2. Goodwin RA Jr, Shapiro JL, Thurman GH, Thurman SS, Des Prez RM. Disseminated histoplasmosis: clinical and pathologic correlations. Medicine (Baltimore). 1980;59(1):1–33.
3. Kumar N, Singh S, Govil S. Adrenal histoplasmosis: clinical presentation and imaging features in nine cases. Abdom Imaging. 2003;28(5):703–8. https://doi.org/10.1007/s00261-003-0010-5.
4. Gupta P, Bhalla A, Sharma R. Bilateral adrenal lesions. J Med Imaging Radiat Oncol. 2012;56(6):636–45. https://doi.org/10.1111/j.1754-9485.2012.02435x. Epub 2012 Sep 27
5. Larbcharoensub N, Boonsakan P, Aroonroch R, Rochanawutanon M, Nitiyanant P, Phongkitkarun S, Poonvutikul S, Watcharananan SP, Ngarmukos C. Adrenal histoplasmosis: a case series and review of the literature. Southeast Asian J Trop Med Public Health. 2011;42(4):920–5.
6. Paolo WF Jr, Nosanchuk JD. Adrenal infections. Int J Infect Dis. 2006;10(5):343–53. https://doi.org/10.1016/j.ijid.2005.08.001. Epub 2006 Feb 17. PMID: 16483815; PMCID: PMC7110804
7. Deepe GS Jr. Immune response to early and late Histoplasma capsulatum infections. Curr Opin Microbiol. 2000;3(4):359–62. https://doi.org/10.1016/s1369-5274(00)00104-1.

A 38-Year-Old Male Presented with Fluctuating Quadriparesis

6

Introduction

Graves' disease often is associated with a variety of other autoimmune diseases as well as several cardiac and neurologic complications. Of the neurologic manifestations, thyrotoxic periodic palsy is a rare channelopathy with dramatic presentation. It is more common in young Chinese males with GD where treatment with ATD and β-blocker is all that is necessary to prevent further episode. Here, we discuss the case of a 38-year-old male with Graves toxicosis presented with abrupt flaccid quadriparesis. He was found to have severe hypokalemia that was complicated by heart block precipitated by β-blocker therapy. It highlights the challenges that we faced and dealt with while evaluating and managing such a patient.

Case Scenario

A 38-year-old garment factory worker was admitted with a complaint of recurrent weakness of all four limbs over the last 2 months. He was diagnosed with Graves thyrotoxicosis based on typical symptoms like palpitation, tremor, heat intolerance, weight loss, and restlessness for 2 years. The symptoms were associated with neck swelling and protrusion of eyeballs. Investigations done at that time showed raised FT3 and FT4, suppressed TSH, high radio-iodine uptake (at 2 h, 31%; at 24 h, 52%), and diffusely increased radiotracer concentration on thyroid scan (Table 6.1). He started on tab. carbimazole 45 mg daily but discontinued the drug by himself after a few months. During the last 2 months, his initial symptoms reappeared, and this time, he also noticed moderate limb weakness. He had difficulties in daily activities like standing from a sitting position, doing overhead work, and easy fatigability. The weakness tends to get aggravated after heavy manual work, dehydration, and following a heavy meal. Each episode persisted for several hours and occurred at least four times in the previous 2 months. He gave no H/O tingling, numbness,

S. Jahan, *Endocrinology Case Studies*,
https://doi.org/10.1007/978-981-95-2770-0_6

Table 6.1 Biochemical report of the patient showing severe toxicosis with a high-uptake thyroid scan

Parameter/date	5.2.17	6.3.19	12.3.19
FT4	**59.3**	**55.6**	–
FT3	**15.9**	**10**	–
TSH	**0.002**	**0.09**	–
RAIU of thyroid	**31% at 2 h** **52% at 24 h**	**24% at 2 h** **11% at 24 h**	–
Thyroid scan	Diffuse, increased radiotracer uptake		–
Serum electrolyte		Na—142, **K—2.5**, Cl—102 mmol/L	Na—144, Cl—101, **K+—1.3**, HCO3—28 mmol/L
ABG		PH—7.45, HCO_3 —23.8 mmol/L	
24 h urinary electrolyte		Na—21 mmol, K—16 mmol/day	
ECG			2nd degree AV block with inferior ischemia

diplopia, dysphagia, breathing difficulty, muscle pain, skin rash, any diurnal variation of weakness, and intake of steroids or other offending drugs. He was a past smoker but did not give H/O alcohol or substance abuse. None of his family members had any similar type of illness.

Q1. What is the most probable diagnosis for limb weakness in this patient?
The most probable diagnosis is thyrotoxic channelopathy.

Q2. What other factors might contribute to weakness in a thyrotoxic patient?
(a) Metabolic myopathy
(b) Cervical myelopathy
(c) Myasthenia Gravis
(d) Familial periodic paralysis
(e) All the above

Considering the scenario, thyroid-associated channelopathy seemed most probable. The patient did not have any family history of such disorder, making the diagnosis of familial periodic palsy less likely. Metabolic myopathy related to toxicosis is of course a contributory factor to be considered as uncontrolled thyrotoxicosis may lead to generalized weakness as well. Autoimmune diseases often tend to coexist. So, Myasthenia Gravis may be a common accompaniment in Graves' disease, as in APS type-2. However, there was no diplopia, ptosis, or diurnal variation of weakness, making this a less likely diagnosis. Cervical myelopathy due to vascular

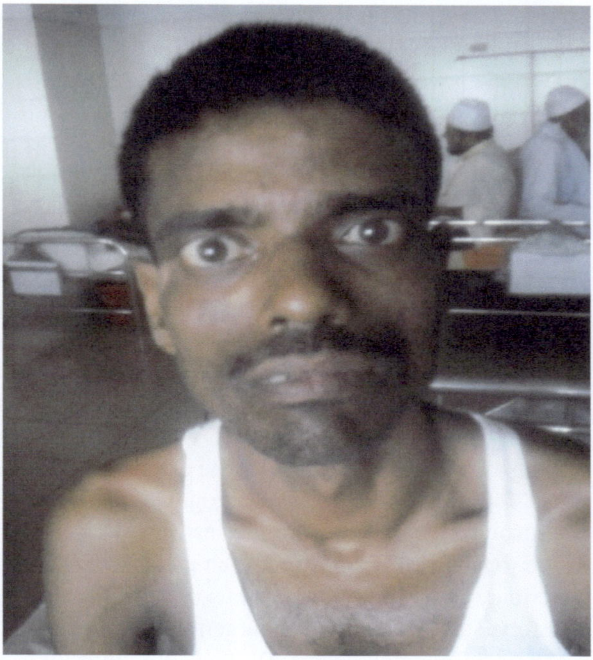

Fig. 6.1 The typical staring look of the emaciated patient. (The photograph was taken with the permission of the patient)

malformation may present with fluctuating weakness. That was a rare possibility here as he had no other symptoms related to myelopathy except weakness.

Examination revealed an emaciated man (BMI–16 kg/m^2) having an anxious, frightened look with warm hands and fine tremors (Fig. 6.1). His pulse rate was 100 beats/min and regular, and BP was 100/70 mm of Hg. Diffusely enlarged, smooth-surfaced, nontender thyroid gland having a firm consistency without retrosternal extension or thyroid bruit was noted. Ophthalmologic examination showed intact visual acuity and light reflex and normal-appearing fundus. There was no diplopia, periorbital edema, or lid lag. However, he had significant exophthalmos (22 mm in the right and 18 mm on the left eye) and keratopathy in the left eye. Nervous system examination revealed decreased muscle bulk in all four limbs; power—4/5 proximally and 5/5 distally; all reflexes were brisk, plantar responses were bilaterally flexor, clonus was absent, and sensory function and cranial nerves were intact.

Q3. What investigations are required at this point?
- (a) Serum electrolyte
- (b) Anti-TPO Ab
- (c) Radio-iodine uptake and thyroid scan
- (d) Repetitive nerve stimulation test (RNS)
- (e) 24 h urinary electrolytes

His FT4 and FT3 were high, TSH was suppressed, and radio-iodine uptake at 2 h was elevated. The antibody specific for GD is TRAb or TSI which was deemed unnecessary as the diagnosis was obvious on other clinical/biochemical parameters. This antibody test is not routinely done and is only required in certain scenarios like euthyroid Graves orbitopathy, neonatal thyrotoxicosis, or unpredictably fluctuating thyrotoxicosis. On the other hand, anti-TPO Ab test is done frequently for the diagnosis of Hashimoto's thyroiditis mostly presenting as primary hypothyroidism. It is also quite common in the general population. RNS is the neurodiagnostic test for Myasthenia Gravis which was not required at this point as it seemed less likely. Serum and 24 h urinary electrolytes need to be measured routinely in thyrotoxic periodic palsy to establish the diagnosis. They were likely to reflect hypokalemia with normal urinary K excretion and normal findings on arterial blood gas analysis (ABG) excluding the renal and nonrenal causes of hypokalemia (renal tubular acidosis, Bartter syndrome, diuretic/laxative abuse, or primary/secondary aldosteronism). Our patient had severe hypokalemia with normal urinary electrolytes and ABG report (Table 6.1).

On the 5th night of admission, he experienced pain and discomfort in the lower limbs, followed by weakness. It progressed to a substantial extent by the morning and involved the upper extremities as well, limiting his mobility. On query, he confessed to having a carbohydrate-rich meal the last evening. On examination, flaccid quadriparesis with diminished muscle power, tone, and jerks were found. Serum electrolyte and ECG were sent urgently. He was found to have severe hypokalemia, bradycardia, and 2nd degree AV block with inferior ischemia.

Q4. What immediate measures are to be taken now?
- (a) I/V bolus potassium
- (b) Oral potassium and I/V potassium infusion
- (c) Omission of propranolol
- (d) Cardiac monitoring
- (e) Increased dose of propranolol

The patient had severe hypokalemia which needed to be corrected immediately. However, the I/V K bolus is completely contraindicated as it may lead to fatal cardiac arrhythmia, severe burning pain, and it also lacks the property of immediate action. So, he was resuscitated with a slow I/V infusion of K in 0.9% N/S. At the same time, oral potassium syrup and dietary supplementation with tender coconut water or banana intake were encouraged. Cardiac consultation was done, and he was advised to have continuous ECG monitoring and omission of propranolol as it might precipitate complete conduction block, and the dose of tab. carbimazole was titrated up to 30 mg/day. The patient gradually improved on this treatment, regaining his muscle power with complete recovery over the next 48 h. Ultimately, he was discharged on tab. carbimazole 30 mg daily. He was also restarted on propranolol (30 mg/day) as his cardiac problem reversed.

Discussion

Graves' disease is one of the most common yet most enigmatic thyroid diseases. It accounts for 50%–80% of all cases of thyrotoxicosis [1]. It can manifest with vivid manifestations, complications, and associations with other autoimmune diseases. Neurologic manifestations include proximal myopathy, tremors, and heightened reflexes. *Thyrotoxic periodic palsy* is a rare channelopathy complicating thyrotoxicosis. It has a prevalence of 0.1–0.2% with >90% of cases occurring in Asians. It occurs in about 10% of young, Chinese thyrotoxic males. It is often precipitated by vigorous exercise, large CHO meal that leads to overactivation of the Na–K ATPase channel. The thyrotoxic patient gives a classic history of periodic weakness leading to flaccid quadriparesis. During the attack, the serum K level is found to be low. However, unlike other etiologies of muscle weakness (myopathy/neuropathy), persistent progressive deficit does not occur, and the patient is often normal in the interictal period. The absence of significant family history and diurnal variability of the weakness rule out the possibility of familial periodic paralysis and Myasthenia Gravis, respectively. This condition may lead to significant paralysis and cardiac arrhythmias. Timely recognition and urgent therapy with potassium supplementation and propranolol are lifesaving, though in our case, it was stopped temporarily to avoid cardiac conduction block. Treatment of thyrotoxicosis with carbimazole along with propranolol is all that is required to prevent a future attack.

Answers

Q2. (a), (c)
Q3. (b), (c), (d)
Q4. (b), (c), (d)

Reference

1. Mandel SJ, et al. Thyrotoxicosis. In: Melmed S, et al., editors. Hormones and disorders of mineral metabolism, William's textbook of endocrinology. 12th ed. Philadelphia: Elsevier Saunders; 2011. p. 566.

A 19-Year-Old Female Presented with Clitoromegaly

7

Introduction

A young female presented with primary amenorrhea, clitoromegaly, and other signs of virilization. These features coupled with her very high plasma androgen level prompted CT scanning of abdomen, revealing bilateral adrenal tumor. She completely returned to a normal state both clinically and biochemically following resection of the tumor. Histopathology was suggestive of pure virilizing benign adrenal adenoma which is a relatively rare entity leading to such presentation.

Case Scenario

A 19-year-old unmarried, female government employee presented with non-establishment of menstruation till date. She also noticed a fleshy growth in introitus in the last 7 years. She was born to a nonconsanguineous marriage; her mother stated an uneventful antenatal period. She could not recall intake of any drug during that period. She was delivered at home by NVD and had age-appropriate developmental milestones. At birth, she was given a female sex assignment and brought up accordingly. There was no history suggestive of a salt-wasting crisis in the immediate neonatal period or thereafter. None of her first degree relatives had similar type of illness, and there was no family history of genital ambiguity, neonatal death, infertility, cryptorchidism or herniorrhaphy at an early age. Ultimately, she was brought to a physician when she failed to establish menstruation even at 19 years of age. She also complained of having a fleshy lump at introitus that progressively enlarged over the last 7 years. Her relatives noticed that she had developed excessive growth of hair in a male pattern in the preceding 4 years. On query, she also admitted of deepening of her voice and poor development of breasts till date.

Physical examination revealed a phenotypic female with masculine body habitus; all her vital parameters were normal, height—149 cm, BMI—20.7 kg/m^2. She

S. Jahan, *Endocrinology Case Studies*,
https://doi.org/10.1007/978-981-95-2770-0_7

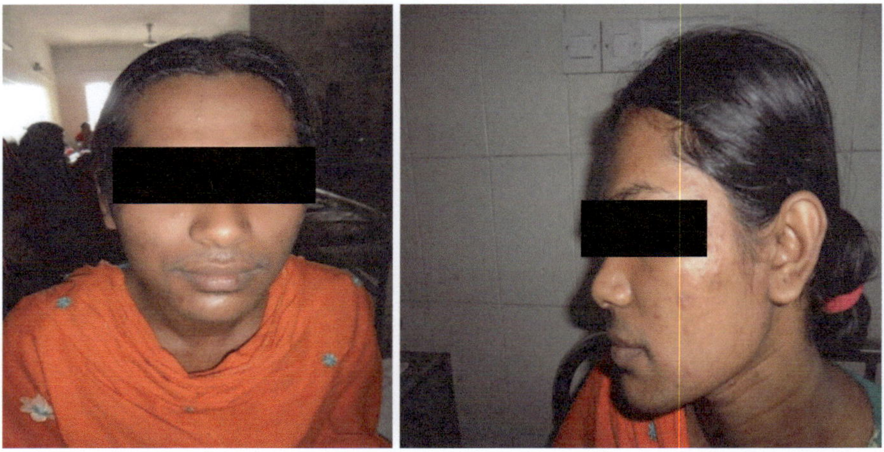

Fig. 7.1 Masculine appearance with hirsutism, acne, and fronto-temporal balding

had signs of marked virilization like frontotemporal balding, husky voice, and severe hirsutism with a Ferriman Gallwey score of 18 (Fig. 7.1). Sexual maturation rating by Tanner staging revealed poor breast development (B2) and abundant pubic hair (P5) in a male pattern of distribution. Examination of genitalia showed the presence of marked clitoromegaly (clitoral index—300 mm^2, length—20 mm, and width—15 mm), vagina could accommodate one finger, and cervix was palpable. There was no palpable lump at or beneath the inguinal region.

Q1. What is the most likely etiology contributing to her physical symptoms?
   (a) Congenital adrenal hyperplasia (CAH)
   (b) Androgen-secreting ovarian/adrenal tumor
   (c) Polycystic ovarian syndrome (PCOS)
   (d) Mullerian agenesis
   (e) 46XY gonadal dysgenesis

The likeliest possibility was an androgen-producing tumor. Virilization in a female is quite different from simple hirsutism because it indicates only a handful of differentials having sinister etiology. These include androgen-secreting tumors, classic simple virilizing CAH, and exposure to androgenic drugs. Other underlying etiologies of hirsutism do not usually lead to frank virilization. The patient neither had any H/O consanguinity, a salt-wasting crisis, or similarly affected siblings, nor did she have any feature of hyperpigmentation, early adrenarche, or height acceleration indicative of CAH. The important clue from history was the onset of clitoromegaly at 12 years with the subsequent appearance of a virilizing feature. If it were due to CAH and clitoromegaly, it would have been present since birth leading to an earlier diagnosis. On the other hand, PCOS leads to oligo-amenorrhea and hirsutism/acne with polycystic ovarian morphology on USG. The onset is in the perimenarche period in a female often having H/O early pubarche. However, clitoromegaly essentially excludes PCOS, though rarely it might be associated with it. Mullerian agenesis popularly known as Myers–Rokitansky–Kuster–Hauser syndrome (MRKHS) can explain primary amenorrhea but cannot account for her virilizing features. 46,XY gonadal dysgenesis like Swyer syndrome may present with female phenotype with sexual infantilism where a microphallus could be confused with an enlarged clitoris in an apparent female infant. However, the relatively recent onset of the disease with rapidly progressive virilization pointed in another direction, namely androgen-producing tumor. While dealing with such a case, drugs should be kept in mind and always enquired of with scrutiny.

Q2. What should be the initial investigation? What other investigations are warranted?
 (a) Karyotyping
 (b) USG of pelvic organs
 (c) HCG stimulation test
 (d) 17-hydroxy-progesterone
 (e) Serum prolactin, IGF-1

In the case of hyperandrogenism, the initial test should be for total testosterone. If it is markedly high, then it indicates either androgen-producing tumor or sometimes CAH (total testosterone >5 nmol/L or > 200 ng/dL). Her total S. testosterone was very high—1115 ng/dL. Karyotyping is an essential investigation that would exclude 46,XY DSD in doubtful cases. Our patient had a typical female karyotype of 46,XX. USG of pelvic organs or TVS in a married lady should be done to see the presence of uterus and adnexa and any androgen-producing ovarian tumor. Her USG of pelvic organs showed a relatively small uterus (3 × 4) cm, multiple tiny follicles in the right ovary, and a small cyst in the enlarged left ovary (cyst size—4.1 × 3.5 × 2.6 mm) (ovarian vol—20.4 cc). Her serum LH, FSH, and estradiol were apparently normal, excluding hypo–/hypergonadotropic hypogonadism or hypothalamic disorder as an underlying cause of amenorrhea. Hyperprolactinemia or acromegaly may cause hirsutism/oligomenorrhea. However, none of them lead to virilization. So, serum prolactin or IGF-1 levels were not necessary here. CAH was effectively ruled out as the report of basal plasma cortisol, serum 17 (OH) progesterone, and plasma ACTH were all within normal limits. Serum 17 (OH) progesterone <2 μg/mL essentially rules out CAH, and in this case, it was only1.1 μg/mL. On the other hand, serum DHEAS—the androgen derived exclusively from adrenals—was found to be significantly high—560.20 μg/dL, favoring an androgen-producing adrenal tumor (Table 7.1). HCG stimulation test is indicated in difficult cases of 46XY DSD to determine the presence of any viable testes. It was not an essential test either considering her female karyotype.

CT scan of the abdomen in the adrenal protocol was done that revealed bilateral well-circumscribed, rounded adrenal masses (5 × 0.3.5 cm) and (4 × 3) cm on the right and left; other abdominal organs were unremarkable (Fig. 7.2). FDG-PET scan showed intense and diffuse uptake in both the adrenals without evidence of any

Table 7.1 Hormonal reports of the patient showing significant biochemical hyperandrogenism

Parameter	Patient's value (before surgery)	Patient's value (after surgery)	Reference range
Serum total testosterone	**1115 ng/dL**	34 ng/dL	63–120 ng/dL
Serum LH	6.15 miu/L	3.5 miu/mL	1.10–11.6 mIU/mL
Serum estradiol	45 pg/mL	65 pg/mL	30–400 pg/mL
Serum FSH	9.76 miu/L	4.6 miu/mL	2.8–11.30 mIU/mL
Basal cortisol	495 nmol/L	11 nmol/L	138–690 nmol/L
Serum 17(OH) progesterone	1.1 μg/mL	–	<2 μg/mL
Serum DHEAS	**560.20 μg/mL**	26 μg/mL	145–395 μg/mL

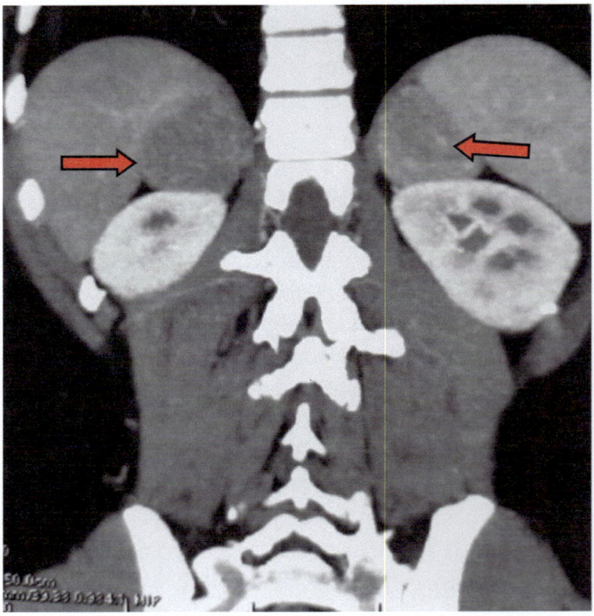

Fig. 7.2 Bilateral adrenal masses in the CT scan

extra-adrenal activity (Fig. 7.3). She was referred to an adrenal surgeon and underwent open bilateral adrenalectomy.

Q3. What are the histopathologic features that should be noted to exclude malignancy in the excised adrenal tumor?
   (a) Abnormal mitotic figure
   (b) Trabecular architecture
   (c) Clear/vacuolated cytoplasm
   (d) Diffuse architecture
   (e) Eosinophilic cytoplasm

Fig. 7.3 FDG-PET showing activity in adrenals

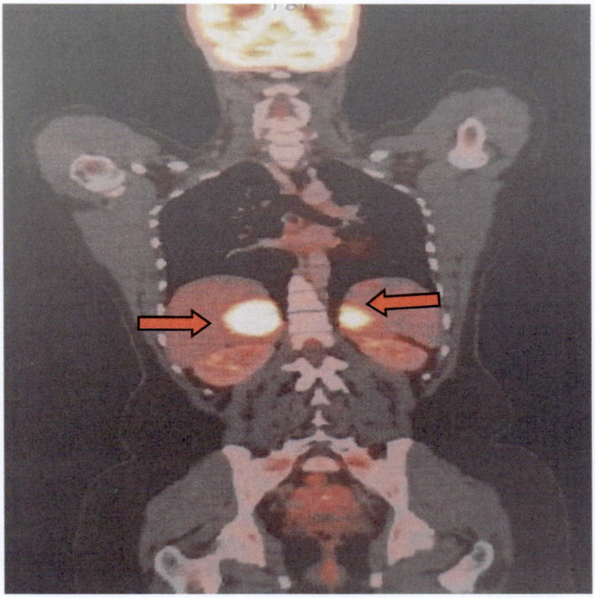

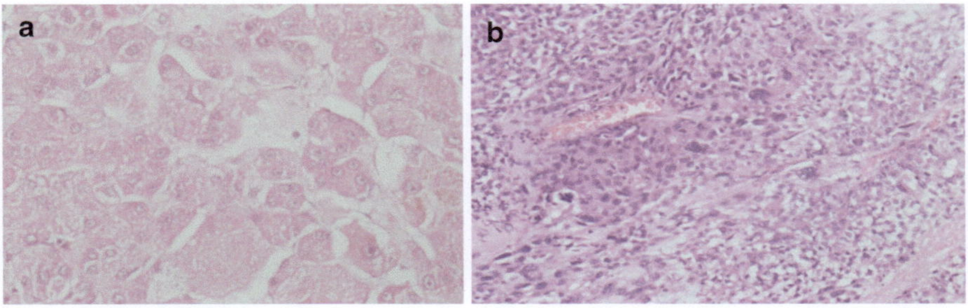

Fig. 7.4 Histopathologic character of benign adrenal adenoma (**a**) and adrenocortical carcinoma (**b**)

It is often difficult to differentiate between benign and malignant endocrine tumors. Though the surest sign of malignancy is distant metastasis, there are certain other features suggestive of malignancy in an adrenal lesion. The commonly used Weiss criteria have got important prognostic as well as diagnostic value in the evaluation of such a tumor. A benign tumor usually has columnar/trabecular architecture consisting of clear/vacuolated cells with no/a few abnormal mitotic figures (<5/HPF). On the other hand, a malignant tumor has a diffuse architecture with eosinophilic cells having an abundant abnormal mitotic figure, and there is often a vascular or capsular invasion as well (Fig. 7.4). The excised tumor in our patient proved to be a purely androgen-secreting benign adrenal adenoma.

Her postoperative period was uneventful, except for the development of adrenal insufficiency for which she was advised lifelong steroid replacement. Immediately after surgery, her hormonal abnormalities reverted to normal, and regular menstruation ensued with the regression of virilization (Table 7.1).

Discussion

The major active androgen in a female is testosterone. Of the total active androgen pool, 25% is secreted directly from the ovaries and another 25% from the adrenals. The rest 50% comes from the peripheral conversion of other androgens (androstenedione, DHEA, and DHEAS). DHEAS is virtually adrenal in origin, whereas the majority of DHEA comes from the adrenals. Testosterone is further activated at the hair follicle by 5α reductase into dihydrotestosterone which leads to hirsutism or virilization if produced in excess. A small amount of testosterone is also converted into estrogen by aromatase in adipose and other tissues. In a female with hirsutism and virilization, if the serum total testosterone is >5 nmol/L, it is indicative of a virilizing tumor, CAH, or exposure to androgenic drugs. In ovarian virilizing tumors, only serum testosterone is remarkably high, whereas in adrenal virilizing tumors, there is also the significant elevation of DHEA and DHEAS (Fig. 7.5).

Androgen-secreting tumors can produce hirsutism and virilization in 90–100% of the patients and amenorrhea in 40–60% [2, 3]. The *virilizing adrenal tumor* is commonly malignant which usually co-secretes multiple hormones. Pure virilizing benign adrenal adenomas are rare. Malignancy likelihood in adrenal tumors has been usually correlated with the tumor size. In this sense, adrenal masses greater

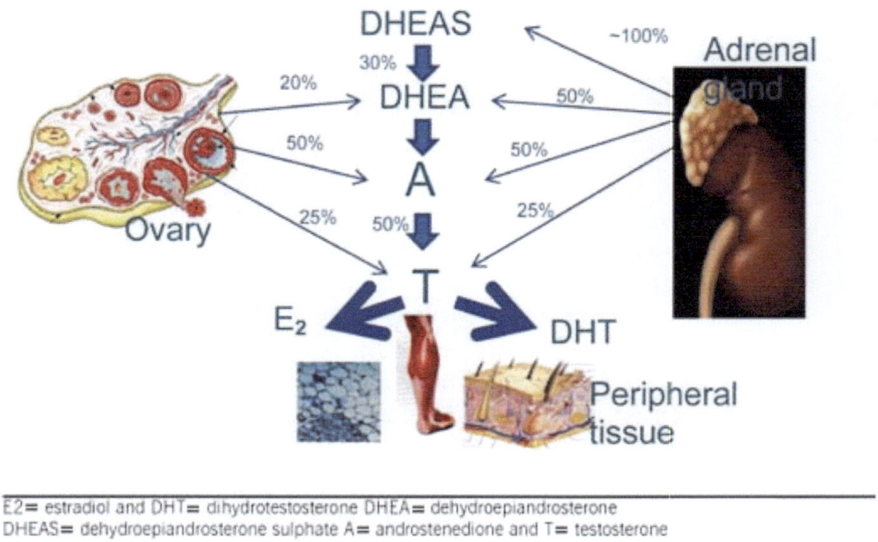

Fig. 7.5 Androgen synthesis pathway in female. (Source: Fonseca et al. [1])

than 4 cm are typically associated with malignancy. Even though it has been described that benign pure androgen-secreting adrenal adenomas could have a size close to 10 cm, most of them are smaller [4]. Malignant tumor has a recent onset and rapidly progressive course in contrast to benign adenoma that has a long and slowly progressive history. The final diagnosis depends on the clinical feature, imaging phenotype, gross appearance, and histopathologic nature of the tumor.

Answers

Q1. (b)
Q2. (a), (b), (d)
Q3. (a), (d), (e)

References

1. Fonseca HP, Scapinelli A, Aoki T, Aldrighi JM. Androgen deficiency in women. Rev Assoc Med Bras. 2010;56(5):579–82.
2. Derksen J, Nagesser SK, Meinders AE, Haak HR, Van De Velde CJH. Identification of virilizing adrenal tumors in hirsute women. N Engl J Med. 1994;331(15):968–73.
3. Cordera F, Grant C, van Heerden J, et al. Androgen-secreting adrenal tumors. Surgery. 2003;134(6):874–80.
4. Moreno S, Montoya G, Armstrong J, et al. Profile and outcome of pure androgen-secreting adrenal tumors in women: experience of 21 cases. Surgery. 2004;136(6):1192–8.

A 40-Year-Old Painter with Weight Loss and Skin Rash

8

Introduction

This case scenario highlights the minor and major side effects of anti-thyroid drugs namely PTU, methimazole, and carbimazole. A middle-aged male with hyperthyroid Graves' disease developed hypersensitivity skin rash following the initiation of carbimazole. The subsequent discussion focuses on how his complication and toxicosis were managed and what cautions are to be exercised while prescribing ATD and specially emphasizes on the management of agranulocytosis.

Case Scenario

A 40- year-old painter presented with unintentional weight loss of about 23 kg within the last 6 months. He had lost his appetite during this period, and his bowel moved four to six times daily with the passage of semisolid stool. He complained of undue fatigability and tiredness. On query, he gave H/O palpitation, restlessness, and tremor along with heat intolerance. Based on a typical clinical feature, he was diagnosed with hyperthyroid Graves' disease. Then, he was started on tab. carbimazole 40 mg/day, and his symptoms immediately improved. However, 2 weeks after the start of treatment, he developed an itchy skin rash over hand and feet. The rashes were initially reddish and subsequently became blackish with desquamation. He stopped the medication by himself 2 weeks after the onset of rashes.

Q1. What important points in history are to be elicited now?

(a) Fever
(b) Sore throat
(c) Cough, chest pain, and sputum production
(d) Drugs like antibiotics and anticonvulsants
(e) Color of urine

S. Jahan, *Endocrinology Case Studies*,
https://doi.org/10.1007/978-981-95-2770-0_8

From history, it was evident that the patient probably had developed erythema multiforme due to drug hypersensitivity. However, it might also occur due to viral or bacterial infections (HSV, mycoplasma, pneumococcus, salmonella, etc.). So, H/O fever, diarrhea, pneumonia, and sore throat should be sought. Medications other than antithyroid drugs (ATD) might also be responsible that has to be kept in mind. Involvement of mucous membranes and blister formation with the subsequent epithelial detachment are the features of a serious disease that need to be looked for immediately.

Q2. What are the common adverse reactions of ATD?
- (a) Agranulocytosis
- (b) Stevens–Johnson syndrome
- (c) Dysgeusia
- (d) ANCA-associated vasculitis
- (e) Nephrotic syndrome.

The most common reactions of ATD are nausea, loss of taste, headache, and hair loss, which are usually self-limiting and do not necessarily require drug withdrawal. The most troublesome in this category is a skin rash, which is commonly urticarial and affects between 1% and 2% of patients. A migratory polyarthritis may occur alone or in association with the rash and resolves within 4 weeks of stopping treatment. Much rarer adverse effects include cholestatic jaundice, vasculitis which may be associated with antineutrophil cytoplasmic antibody, a lupus-like syndrome, and nephrotic syndrome.

Q3. What are the life-threatening adverse reactions of ATD?
- (a) Fulminant hepatitis
- (b) Cholestatic jaundice
- (c) Agranulocytosis
- (d) Acute renal shutdown
- (e) Lactic acidosis

Agranulocytosis is the most dreaded complication of ATD, and it occurs in 0.1–0.3% of patients. It necessitates immediate discontinuation of the drug, otherwise it may lead to life-threatening neutropenic sepsis. PTU is the third most common cause of drug-related hepatic failure, whereas methimazole leads to relatively self-limited cholestatic jaundice. On the other hand, ATD is not associated with renal failure or lactic acidosis, which occurs with other common drugs like NSAID.

On admission, he was found to be anxious having a staring look. His BMI was 17.09 kg/m^2, pulse was 100/min, regular, and the recorded BP was 120/70 mm Hg. The thyroid gland was diffusely enlarged, firm, nontender, and freely mobile with no retrosternal extension or thyroid bruit. There was no lymphadenopathy, hepatosplenomegaly, or bony tenderness. Inactive Graves ophthalmopathy was present. There were multiple skin rashes of variable size and shape involving the dorsum aspect of the hand and feet. Some were reddish and edematous, while others showed

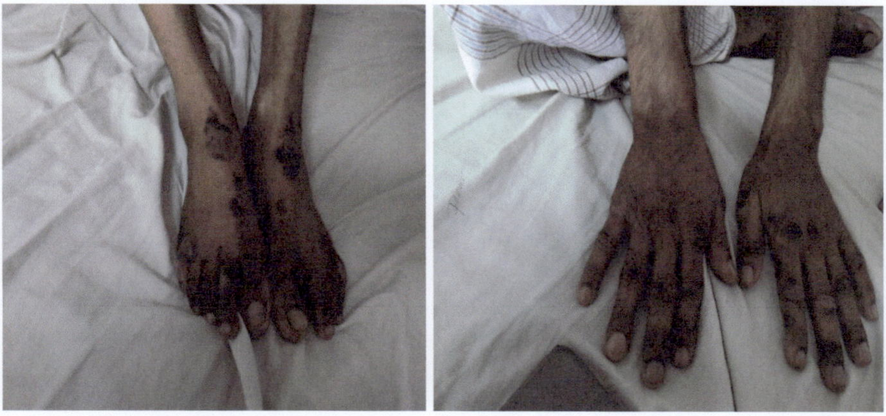

Fig. 8.1 Itchy desquamative rash over the hand and feet of the patient

Table 8.1 Biochemical report of the patient showing severe toxicosis and diffuse uptake in thyroid scan

Parameter	Patient's value	Reference range
CBC with ESR	Hb—14.8%, TC of WBC—7600/cu mm, N—54%, L—33% ESR—34 mm in first hour	Hb—15 ± 2 g/dL, TC of WBC—7 ± 3 × 10^9/l ESR—0–10 mm in first hour
FT4	**6.09**	0.8–1.8 ng/dL
FT3	**17.6**	1.8–4.7 pg/mL
TSH	**<0.004 mIU/mL**	0.35–5.5 μIU/mL
RBS	6.5 mmol/L	–
SGPT	40 iu/L	14–63 IU/L
RAIU of thyroid	**At 2 h—8%** **At 24 h—44%**	At 2 h—4–10% At 24 h—10–25%

blackish pigmentation with desquamation (Fig. 8.1). There was no evidence of an oral ulcer or any mucous membrane involvement. Investigations showed high FT4 and FT3 with suppressed TSH and high radio-iodine uptake at 24 h. Thyroid scan showed hyperfunctioning status, and USG of the neck revealed diffuse thyromegaly with uniform echo texture of both lobes except for a hypoechoic cystic nodular area at the lower pole of the right lobe (Table 8.1).

He was treated with tab. propranolol 20 mg tds and tab. phenobarbital 30 mg OD to relieve toxic symptoms, and tab carbimazole was given at a low dose. Skin lesion was treated with antihistamine tablets and local steroid and emollient.

Q4. How should he be treated for toxicosis in the future?

(a) ATD in full dose
(b) Radio-iodine ablation
(c) Thyroidectomy

(d) Lithium carbonate
(e) Lugol's iodide

The patient experienced a minor side effect of ATD that is hypersensitivity skin rash. In this case, reinstitution of the drug after resolution of the skin rash is recommended. However, if the patient had experienced a life-threatening side effect of ATD like agranulocytosis, then re-initiation of ATD is contraindicated. In these cases, surgery or radio-iodine ablation should be offered once the patient is euthyroid. Lithium carbonate is used as an adjunct to radio-iodine therapy when the patient has allergic reactions or is intolerant to ATD. It inhibits thyroid hormone release. Lugol's iodide is used in thyroid storms, before thyroid surgery or severe thyrocardiac disease or in cases of ATD intolerance and allergy prior to definitive therapy. ATD in full dose was gradually started in our patient without further adverse events. He became euthyroid within a few months, and the drug could be eventually withdrawn after 2 years.

Discussion

The major class of drugs used in thyrotoxicosis is *thionamide,* namely PTU, methimazole, and carbimazole. They are relatively safe with adverse reactions occurring only in a handful of cases. Mild side effects occur in up to 20% of patients and include skin reactions, arthralgias, gastrointestinal symptoms, an abnormal sense of taste, and, occasionally, sialadenitis. Other effects include thrombocytopenia, enlargement of lymph nodes, a lupus-like syndrome including the development of antineutrophil cytoplasmic antibody (ANCA)-positive vasculitis, and toxic psychoses. The mechanisms underlying these reactions are not known, although some reactions disappear with the continuance of treatment. Among the more serious side effects, the one that elicits the most concern is agranulocytosis, which occurs in 0.1% to 0.3% of patients, usually within the first few weeks or months of treatment. Older patients and those taking higher antithyroid doses are at a higher risk. Propylthiouracil has, in rare instances, been associated with fulminant hepatic necrosis, and it is the third most common cause of drug-related liver failure, accounting for 10% of all drug-related liver transplantations. Cessation of the treatment results in recovery in most cases. This propylthiouracil-associated liver failure may occur at any time during therapy, so routine monitoring of liver function may not be helpful. Methimazole is associated with cholestasis rather than hepatic necrosis, and the risk is increased with higher doses administered in older patients. There are no reported cases of liver transplantation attributed to methimazole toxicity. In the case of minor side effects like hypersensitivity skin rash, the drug can be restarted after successful treatment of the side effects. However, in the case of a major adverse event, re-initiation of the drug is contraindicated, and an alternate form of therapy (RAIU ablation/surgery) should be offered.

Answers

Q1. (a), (b), (c), (d)
Q2. (b), (c)
Q3. (a), (c)
Q4. (a)

Reference

1. Mandel SJ, et al. Thyrotoxicosis. (2011). Hormones and disorders of mineral metabolism. In: Melmed S, et al., editors. William's textbook of endocrinology. 12th ed. John F. Kennedy Blvd. Philadelphia/Elsevier Saunders; 2011. p. 378–81.

A 25-Year-Old Male with Recurrent Panic Attack and Bumpy Lips

9

Introduction

A young normotensive male presented with erectile dysfunction and primary subfertility. The incidental detection of bumpy lips led to the query of "Pheo spell" that was typically present. These symptoms coupled with the presence of mucosal neuroma prompted the evaluation for MEN 2B. He was found to have thyroid medullary carcinoma and pheochromocytoma at the same time. Interestingly his symptoms improved following the resection of tumor, and the research of review literature suggested that erectile dysfunction could be a presenting feature of pheochromocytoma.

Case Scenario

A 25-year-old male presented with the inability to procreate a baby during the 3 years of his marriage. He had erectile dysfunction that was treated with penile prosthesis 4 years back. He had intact libido, normal shaving frequency, and otherwise harmonious conjugal life. He also admitted of having frequent spells of palpitation, headache, and sweating for the last 2 years. He stated that each episode started with chest discomfort, followed by a severe throbbing frontal headache associated with palpitations but not with nausea or vomiting. It was precipitated by stress, exhaustion, and tends to be relieved by taking rest. Each episode persisted for 5–15 min and subsided with profuse sweating. There was no diurnal variation or any aggravating factors like coughing, sneezing, bending forward, etc. He was a smoker consuming 5 packs/year and nonalcoholic. None of his family members suffered from a similar type of illness.

S. Jahan, *Endocrinology Case Studies*,
https://doi.org/10.1007/978-981-95-2770-0_9

Q1. What is the most probable diagnosis accounting for episodic headache?
  (a) Migraine headache
  (b) Pheochromocytoma
  (c) Severe anxiety and panic attack
  (d) Paroxysmal arrhythmia
  (e) Thyrotoxicosis

Considering history, all the above are possibilities that need to be scrutinized. The headache could be migrainous though it was not unilateral and not associated with nausea, vomiting, photophobia, or phonophobia. It always gave way to profuse sweating that is also atypical for migraines. On the other hand, there was no situational clue leading to severe anxiety or panic attacks that lasted so briefly. He gave no history of heat intolerance, weight loss, excessive sweating, tremors, hyperdefecation, or a goiter, making the likelihood of thyrotoxicosis negligible. There was no history of dyspnea, chest pain, or any missing beat suggestive of cardiac arrhythmia where a severe headache was implausible. The episode was typical of hyperadrenergic or "Pheo spells" consisting of headache, palpitations, and sweating, but the absence of fluctuating HTN made it less likely.

Q2. Which of the following statement regarding pheochromocytoma is true?
  (a) Pheochromocytoma is always characterized by fluctuating HTN.
  (b) Persistent HTN is less common than fluctuating HTN in Pheo.
  (c) Hypotension or normotension effectively rules out Pheo.
  (d) HTN with hypokalemia is typical of Pheo.
  (e) None of the above.

Pheochromocytoma is not always associated with paroxysmal HTN, and persistent rather than fluctuating HTN is more common. In 50% of cases of pheochromocytoma, HTN is persistent, and in 30% of cases, it is paroxysmal. At the same time, 20% of patients are normotensive, and it is not uncommon for a hypotensive spell to occur after a hyperadrenergic episode or at other times. Patients whose pheochromocytoma is diagnosed in the presymptomatic stage may have normal blood pressure. The lability in blood pressure can be attributed to the episodic release of catecholamines, chronic volume depletion, and impaired sympathetic reflexes. In addition to volume depletion, altered sympathetic vascular regulation may have a role in orthostasis, which may be observed in patients with pheochromocytoma. Symptoms of orthostatic hypotension (e.g., lightheadedness, presyncope, and syncope) may dominate the presentation, especially in patients with epinephrine- or dopamine-predominant tumors (1).

On examination, he was found to be anxious and cooperative. He had an elongated face with thick, bumpy lips. There were multiple soft papulonodular lesions over the tip and lateral aspects of the tongue and lips with malalignment of teeth (Figs. 9.1 and 9.2). His height was 177 cm, weight: 59 kg, BMI: 18.62 kg/m^2, arm span: 180 cm, and upper segment: lower segment = 0.95:1. His pulse rate was 80/min, regular, and BP was 100/70 mm Hg (lying) and 90/70 mm Hg (standing).

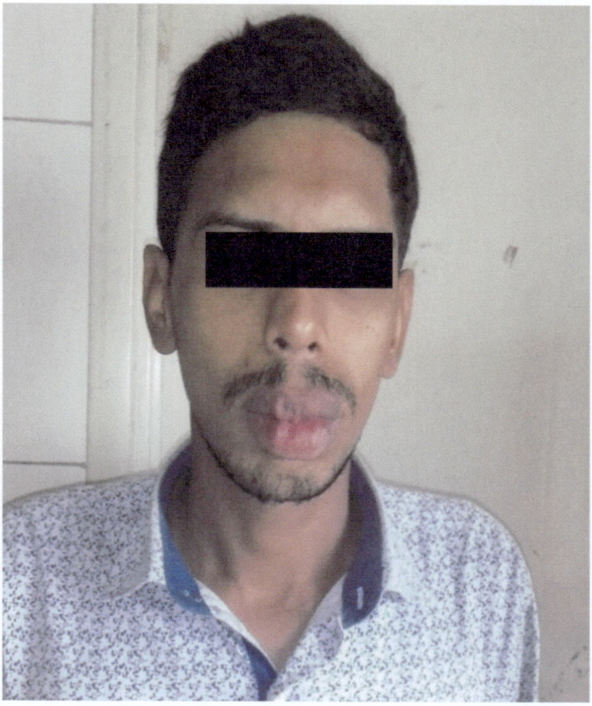

Fig. 9.1 Thick, bumpy lips of the patient

There was no lymphadenopathy or thyromegaly. His secondary sex characteristics were normal, though testicular volume was relatively small (right—12 mL, left—10 mL). His family history was unrevealing except for the recall of thick, bumpy lips that his grandfather had.

Q3. What is the probable diagnosis based on the clinical feature?
 (a) Neurofibromatosis type 1
 (b) Multiple endocrine neoplasia type 1(MEN 1)
 (c) Multiple endocrine neoplasia type 2A (MEN 2 A)
 (d) Multiple endocrine neoplasia type 2B (MEN 2B)
 (e) Isolated pheochromocytoma

Tumor of the supporting element of a nerve fiber or neurofibromatosis is characterized by mucocutaneous neurofibroma, café-au-lait spot, axillary freckles, Lisch nodule in the iris, hearing impairment, or scoliosis in an individual with autosomal dominant inheritance. None were present except mucosal neuroma, making the diagnosis unlikely. MEN 1 is an autosomal recessive disorder consisting of multiple endocrine tumors originating from the pituitary, parathyroid, and pancreas. The commonest variety of pituitary tumors in MEN 1 secrete prolactin or growth hormone, and gastrinoma and insulinoma are the typical pancreatic tumors. Our patient did not have any history of acral enlargement, hypogonadism, or headache

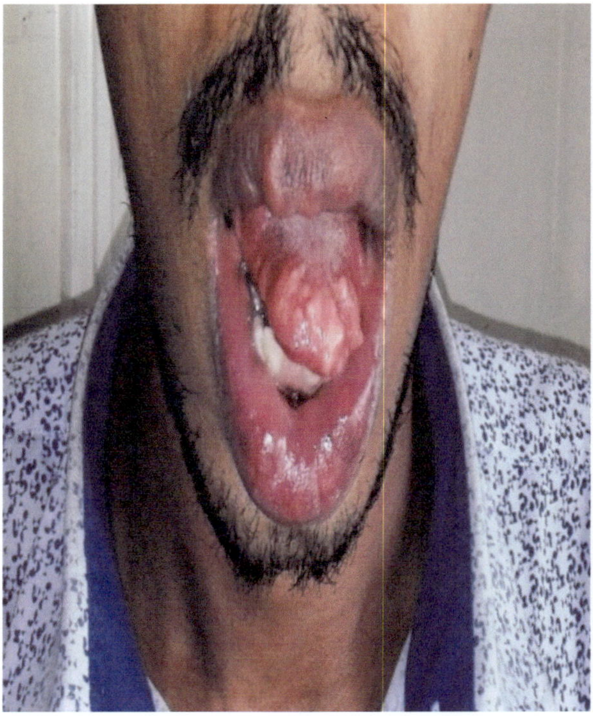

Fig. 9.2 Multiple mucosal neuromas in the patient

suggestive of pituitary tumor or body aches, bone pain, and renal calculi indicative of hyperparathyroidism or multiple GIT ulceration or spontaneous hypoglycemia arising from gastrointestinal NET tumor excluding the possibility of MEN 1. The presence of a typical hyperadrenergic spell along with the characteristic mucosal neuroma suggested MEN 2B which shares the common feature of thyroid medullary carcinoma and pheochromocytoma with MEN 2A. However, hyperparathyroidism is common in MEN 2A, whereas mucosal neuroma and marfanoid habitus occur exclusively in MEN 2B.

Q4. What biochemical investigations were warranted?
- (a) Serum calcitonin
- (b) Calcium panel
- (c) 24 h urinary metanephrine/normetanephrine
- (d) Serum prolactin, IGF-1
- (e) Serum thyroglobulin

He was investigated accordingly after admission. Routine biochemical investigations including the serum calcium panel were normal. Serum calcitonin was found to be exceedingly high: 1755 pg/mL (normal: up to 18.2 pg/dL). Twenty-four hour urinary metanephrine and nor-metanephrine were at the upper limit of normal (Table 9.1). Other investigations like serum prolactin and IGF-1 are required for the

Table 9.1 Biochemical investigations of the patient showing high plasma calcitonin and urinary catecholamine metabolite

Parameter	Patient's value	Reference range
CBC and ESR	Hb:15.7 g/dL, TLC:7.00 × 10^3 /μL, TPC: 320 × 10^3 /μL, ESR: 20 mm at the end of first hour	Hb—15 ± 2 g/dL, TLC—7 ± 3 × $10^{9/l,}$ TPC—150–450 × 10^3/μL ESR—0–10 mm in first hour
OGTT	FPG: 5.1 mmol/L 2-h PG: 3.3 mmol/L	FPG: ≤5.5 mmol/L 2-h PG: ≤7.8 mmol/L
Serum creatinine	0.82 mg/dL	0.5–1.3 mg/dL
Calcium panel	Serum calcium: 9.6 mg/dL, Serum albumin: 45 g/L Serum inorganic phosphate: 3.6 mg/dL, I PTH: 35.1 pg/mL Alkaline phosphatase: 76 U/L	Serum calcium: 9–11 mg/dL, Serum albumin: 34–54 g/L Serum inorganic phosphate: 3–4.5 mg/dL, iPTH: 11–67 pg/mL
TFT	TSH: 1.37 μIU/mL Free T4: 1.24 ng/dL	TSH—0.35–5.5 μIU/mL FT4—0.8–1.8 ng/dL
Serum LH	2.3 mIU/mL	0.8–7.8 mIU/mL
Serum FSH	3.3 mIU/mL	0.7–11.1 mIU/mL
Plasma testosterone	560 ng/dL	270–1734 ng/dL
Serum calcitonin	**1755 pg/mL**	Upto 18.2 pg/dL
24 h urinary metanephrine	**1482.64 nmol/day**	<1775 nmol/day
24 h urinary nor-metanephrine	**369 mcg/day**	75–375 mcg/day

diagnosis of MEN 1, and thyroglobulin is a marker of thyroid follicular/papillary carcinoma. So, these investigations were not necessary for our patient.

Q5. What imaging studies should be done?
 (a) Pituitary MRI
 (b) USG of neck
 (c) Endoscopic USG of pancreas
 (d) Adrenal CT scan
 (e) Octreoscan

The USG of the neck showed multiple tiny thyroid nodules, and a CT scan of abdomen with an adrenal protocol showed a right adrenal mass. The imaging phenotype of the adrenal mass was suggestive of pheochromocytoma [size (2.5 × 3.1) cm, round and regular, hypodense having a precontrast lipid attenuation value of 45 HU, and delayed contrast washout (RWR- < 40%)] (Fig. 9.3). Pituitary MRI was unnecessary as MEN 1 was a rare possibility. At the same time, endoscopic USG or octreoscan was not indicated as pancreatic neuroendocrine tumor also was less probable in this case. Ultimately, after appropriate preparation, he was referred to unilateral adrenalectomy with a plan for total thyroidectomy and family screening.

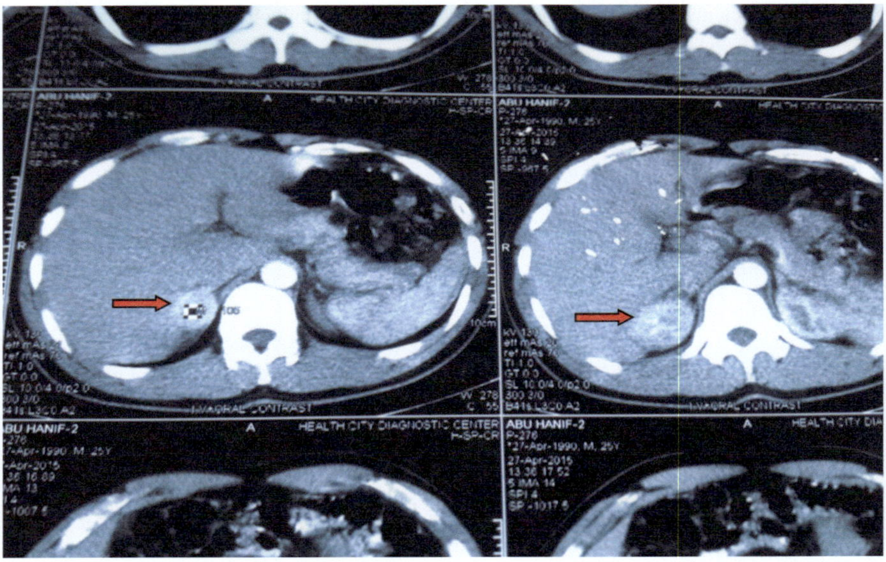

Fig. 9.3 CT scan of abdomen showing the right adrenal mass

Interestingly his erectile dysfunction improved significantly following the removal of the adrenal tumor. It is postulated that high circulating noradrenaline in pheochromocytoma leads to unopposed vasoconstriction of penile vessels and corpora cavernosa leading to erectile dysfunction. Erectile dysfunction can be the presenting symptom of pheochromocytoma as reported by various authors [1].

Discussion

Multiple endocrine neoplasias or multiple endocrine adenomatosis syndromes are named so because they cause tumors in two or more types of hormone-secreting organs and consequently produce syndromes of multiple hormone excess. *MEN 2B* accounts for 5% of patients with MEN 2. MTC often presents in infancy and is highly aggressive. Approximately 75% of patients with MEN 2B are sporadic and present as de novo RET mutations [2]. In addition to pheochromocytomas, which develop in 50% of cases, patients have a characteristic physical appearance with typical facies, ophthalmologic abnormalities (thickened and everted eyelids, mild ptosis, and prominent corneal nerves), and skeletal malformations (marfanoid body habitus, narrow long facies, pes cavus, pectus excavatum, high-arched palate, scoliosis, and slipped capital femoral epiphysis). Patients also have generalized ganglioneuromas throughout the aerodigestive tract and are almost always symptomatic with intermittent diarrhea and constipation. Some patients require surgery early in life because of the gastrointestinal malfunction [3, 4]. It is important to establish the diagnosis of MEN 2B at an early age when there is a possibility that thyroidectomy will be curative (Fig. 9.4).

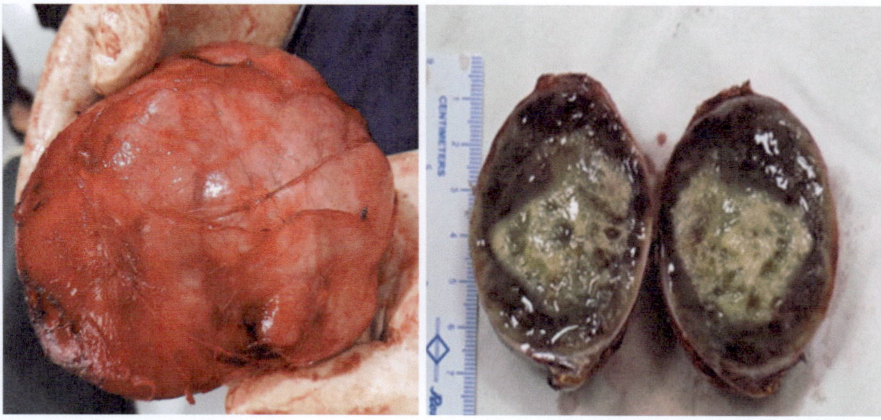

Fig. 9.4 Gross appearance of pheochromocytoma with areas of hemorrhage and necrosis

Answers

Q1. (b)
Q2. (e)
Q3. (d)
Q4. (a), (c)
Q5. (b), (d)

References

1. van den Berge K, van Guldner C. Erectile dysfunction as the presenting symptom of a pheochromocytoma. EJIM. 2007;18:593–4.
2. Yamaguchi K, Abe K, Adachi I, et al. Concomitant production of immunoreactive gastrin-releasing peptide and calcitonin in medullary carcinoma of the thyroid. Metabolism. 1984;33:724–7.
3. Baylin SB, Beaven MA, Buja LM, et al. Histaminase activity: a biochemical marker for medullary carcinoma of the thyroid. Am J Med. 1972;53:723–33.
4. Melvin KE, Miller HH, Tashjian AH Jr. Early diagnosis of medullary carcinoma of the thyroid gland by means of calcitonin assay. N Engl J Med. 1971;285:1115–20.

12- and 8-Year-Old Siblings with Extreme Short Stature 10

Introduction

Two intelligent and cheerful siblings born to a consanguineous parent presented with facial dysmorphism and extreme short stature. They had elevated GH level on repeated sampling in the backdrop of clinical features of GH deficiency or panhypopituitarism. Finally, they were diagnosed of having GH insensitivity syndrome popularly known as Laron syndrome when their IGF-1 and IGFBP-3 level were found to be low. Their condition remained unchanged as recombinant IGF-1 therapy could not be offered.

Case Scenario

A 12-year-old girl and her 8-year-old brother born to a consanguineous parent presented with severe short stature. Their intrauterine periods were uneventful, and they were delivered full-term via vaginal delivery without any complication. Her birth weight was approximately 3 kg as per her mother's recall. The girl was apparently alright till 2 years of age with normal developmental milestones but failed to develop properly thereafter. Her height was almost constant over the last 9 years as compared to her peers (Fig. 10.1 left). The parents noticed their son also facing the same problem since he was 2 years of age (Fig. 10.1 right). Their secondary sexual characteristics were also underdeveloped, and there was no change in clothing and shoe size for the last several years. They had two siblings, a 19-year-old elder brother of average height and a deceased sister who was normal. A paternal aunt and an uncle who died at an age of 4 years and 2 years, respectively, were reported to have short stature.

S. Jahan, *Endocrinology Case Studies*,
https://doi.org/10.1007/978-981-95-2770-0_10

Fig. 10.1 Comparison of growth in siblings with age-matched peers

Q1. What other features should be noted from their birth history?
   (a) Neonatal hypoglycemia and prolonged jaundice
   (b) Microphallus
   (c) Ambiguous genitalia
   (d) Salt-wasting crisis
   (e) Midline cranial defect

The siblings presented with severe short stature that might be due to GH deficiency or panhypopituitarism characterized by prolonged neonatal jaundice and hypoglycemia resulting from GH or cortisol deficiency. Microphallus and midline defects are also common in such cases. Genital ambiguity occurs in disordered sexual differentiation (DSD) that commonly is not associated with short stature. Congenital adrenal hyperplasia (CAH), the commonest cause of 46XX DSD, leads to salt-wasting crisis and growth acceleration rather than deceleration.

The siblings had typical dysmorphic facies characterized by frontal bossing, shallow orbit with the blue sclera, saddle nose, mid-face hypoplasia, thin and silky hair, and pale texture of the skin. In addition, the boy had dental caries and bilateral undescended testes with microphallus. Their intelligence seemed to be normal, and they were very cheerful. The height of the boy was 78 cm, weight—17.5 kg, and body mass index (BMI)—27.68 kg/m^2 with an increased upper-to-lower segment ratio of 1.16, which were 85 cm, 20 kg, 28.76 kg/m^2, and 1.02, respectively, in the case of the girl. When plotted on age- and sex-appropriate growth chart, the height and weight of both the siblings were well below the first percentile; the calculated mid-parental height for the boy was 163 cm, whereas for the girl, it was 150 cm. The sexual maturation rating of the girl by Tanner staging revealed her to be at stage 1 (pre-pubertal).

Q2. What is the possible diagnosis?
- (a) GH deficiency syndrome
- (b) Panhypopituitarism
- (c) Constitutional delay of growth and maturation (CDGM)
- (d) Genetic/familial short stature
- (e) Primary hypothyroidism

The siblings had an appropriate developmental milestone with normal intelligence, and the absence of such stigmata like coarse skin, constipation, puffy face, hypotonia, or slowness of activities made hypothyroidism a rare possibility. There was no family history of short stature or delayed puberty, and both parents were of average height, thereby excluding familial or constitutional etiology. Extreme short stature with delayed puberty could either be due to GH deficiency or panhypopituitarism.

Routine hematologic and biochemical profiles were unremarkable, thereby excluding any systemic illness. Bone age was markedly delayed in both cases, around 2 years and 6–7 years for the boy and the girl, respectively. Karyotyping was as expected, 46XY in the case of the boy and 46XX for the girl. An oral glucose tolerance test (OGTT) was performed, and interestingly three out of four values were found to be in the hypoglycemic range (fasting plasma glucose, FPG—2.4 mmol/L and 3.8 mmol/L; 2 h post glucose—3.4 mmol/L and 4 mmol/L for the brother and sister, respectively). Hormonal assay for anterior pituitary function revealed both to be euthyroid with normal plasma ACTH/basal serum cortisol and prolactin level.

Q3. What other investigations need to be carried out now?
- (a) GnRH stimulation test
- (b) HCG stimulation test
- (c) Co-Syntropin stimulation test
- (d) GH stimulation test
- (e) IGF-1 level

The GnRH stimulation test is a dynamic test designed to assess the activity of the hypothalamic–pituitary–gonadal axis. The girl had a pubertal response on the GnRH

Table 10.1 Hormonal profile of the siblings showing high GH and low IGF-1/IGFBP-3

Hormonal profile	Boy	Girl	Reference level
T4	8.75	11.89	4.5–12 μgm/dl
TSH	2.37	2.28	0.47–5.01 μIU/ml
Basal cortisol	21.0	26.10	3.7–19.4 μgm/dl
Prolactin	6.94	7.39	2.5–17 ng/ml
LH (basal)	_	1.70	0.31–5.29 IU/l
FSH (basal)	_	11.5	0.4–6.5 IU/l
LH (post-GnRH)	_	10.95	–
FSH (post-GnRH)	_	16.82	–
Basal GH	**20.57** **38.10 (Repeat assay)**	**31.89** **7.94 (Repeat assay)**	0.03–14.90 ng/ml
IGF-1	**<25**	**<25**	64–345 μgm/ml
IGFBP-3	**<0.500**	**0.51**	1.6–6.5 μgm/ml

stimulation test (Table 10.1). Co-Syntropin/ACTH stimulation test was not required in the siblings as their basal cortisol was normal, reflecting the adequacy of the HPA axis. HCG stimulation test is sometimes performed in cases of 46XY DSD for the localization of viable testicular tissue or categorizing the disorder. This test was not necessary here.

Surprisingly, their random plasma growth hormone (GH) level was found to be remarkably high rather than low (20.57 ng/mL and 31.89 ng/mL for the boy and the girl, respectively). The GH levels were rechecked at another reliable, specialized lab, and similar reports were obtained (Table 10.1). Typical clinical features of GH deficiency in the backdrop of elevated plasma GH level pointed toward the diagnosis of GH insensitivity syndrome, popularly known as Laron syndrome.

Q4. What are the confirmatory tests needed at this point?
   (a) IGF-1
   (b) Dynamic pituitary MRI
   (c) Kisspeptin
   (d) Ghrelin
   (e) IGFBP-3

Kisspeptin is a neuropeptide that is implicated in the initiation of puberty via its interaction with the GnRH pulse generator. As the boy was in pre-pubertal age and his sister had a pubertal response on GnRH stimulation test, measuring plasma kisspeptin was not necessary. Ghrelin is the hormone secreted from the stomach that stimulates appetite. It has a minor role in inhibiting GH secretion that is not of much clinical relevance. The mediator of GH action is IGF-1 which binds to various proteins in the circulation like IGFBP3. The low level of these molecules when combined with a high level of GH is considered to be confirmatory in GH resistance syndrome. Serum IGF-1 and IGFBP-3 levels were assayed at a pediatric reference lab, and both the results were found to be low as was expected (Table 10.1). Finally, an MRI of the pituitary was done; it showed partial empty Sella in both cases, which was probably an incidental finding. As genetic testing was not locally available, it could not be performed.

Q5. What is the treatment of choice for Laron syndrome?
(a) Recombinant GH
(b) Recombinant IGF-1
(c) Oxandrolone
(d) Recombinant IGF-2
(e) GnRH analog

GnRH analogs, e.g., triptorelin and buserelin, are used in the treatment of central precocious puberty. Oxandrolone, a weak androgen, is sometimes given in the case of short stature to augment pubertal growth spurt. Recombinant GH is the treatment of choice for GH deficiency. It is ineffective in GH resistance syndrome. So, finally, the recommended treatment in such cases is recombinant IGF-1, which could not be offered as it was not commercially available locally.

Discussion

Primary growth hormone resistance or *growth hormone insensitivity syndrome (GHIS)*, also known as *Laron syndrome*, is a hereditary disease caused by deletions or different types of mutations in the growth hormone receptor gene or by post-receptor defects. This disorder is characterized by a clinical appearance of severe growth hormone deficiency with high levels of circulating growth hormone in contrast to low serum insulin-like growth factor 1 values [1]. Clinically, patients with GHIS present in a manner virtually indistinguishable from those with severe GH deficiency. Congenital malformations, craniofacial abnormalities, and other physical features may be noted at birth. Facial bone growth is particularly retarded, and fontanelle closure is delayed, leading to a disproportionate cephalon–facial relationship because of the decreased vertical dimension of the face, with frontal bossing, a saddle nose, shallow orbits, and the setting sun sign of the eyes. Blue sclera may be noted, particularly in patients of Mediterranean or Middle Eastern origin. Pubertal development is delayed, and the pubertal growth spurt is absent, but adult sexual maturation is eventually achieved. Walking and other gross motor developmental milestones are delayed because of the underdeveloped musculature. The hands and feet are small (acromicria) [2]. The cardinal biochemical features of GHIS are low levels of all GH-dependent proteins, including very low or even undetectable serum IGF-1 levels, IGFBP-3, and ALS in association with normal or increased GH levels. The regulation of GH secretion and feedback mechanisms is normal. The most important functional test for the diagnosis is the IGF-1 generation test because serum IGF-1 levels are low and do not increase with the administration of exogenous recombinant GH for days or weeks, demonstrating the state of GH resistance in these patients [2]. The only effective treatment is the daily administration of rIGF-1 starting from early childhood and probably throughout life. The rIGF-1 treatment accelerates linear growth velocity, and appropriate dose titration results in a tripling of the baseline growth rate during the first year of treatment [3].

Answers

Q1. (a), (b), (d)
Q2. (a), (b)
Q3. (a), (d), (e)
Q4. (a), (c)
Q5. (b)

References

1. Cotta OR, Santarpia L, Curtò L, Aimaretti G, Corneli G, Trimarchi F, Cannavò S. Primary growth hormone insensitivity (Laron syndrome) and acquired hypothyroidism: a case report. J Med Case Rep. 2011;5:301.
2. Laron Z. Laron syndrome (primary growth hormone resistance or insensitivity): the personal experience 1958-2003. J Clin Endocrinol Metab. 2004;89(3):1031–44.
3. Chernausek SD, Backeljauw PF, Frane J, Kuntze J, Underwood LE, GH Insensitivity Syndrome Collaborative Group. Long-term treatment with recombinant insulin-like growth factor (IGF)-I in children with severe IGF-I deficiency due to growth hormone insensitivity. J Clin Endocrinol Metab. 2007;92(3):902–10.

A 28-Year-Old Female Presented with Weight Gain and Proximal Myopathy

11

Introduction

This case describes the unique co-occurrence of a pituitary microadenoma and an adrenal nodule in a young female with Cushing syndrome. This young lady with typical stigmata of Cushing syndrome was biochemically confirmed to have an ACTH-dependent disease. However, the presence of an adrenal nodule led to a diagnostic dilemma. The resected pituitary tumor showed Crooke's hyaline change and had positive ACTH immunostaining. Finally, significant clinical and biochemical improvement along with regression of the adrenal nodule following pituitary adenomectomy proved the ultimate diagnosis of Cushing disease.

Case Scenario

A 28-year-old female presented with progressive weight gain over the last 3 months. It was unrelated to her dietary habit or level of physical activity. She noticed that her torso was disproportionately bulky in contrast to thin limbs. During this period, she also developed weakness of both lower limbs manifested by difficulty in standing from a sitting position. Weakness was not related to exercise; there was no diurnal variation, muscle pain, skin rash, dysphagia, or any breathing difficulty. She also complained of persistent bands like headache of mild to moderate severity not associated with nausea, photophobia, or any premonitory symptoms. Recently, she had been found to have uncontrolled HTN requiring multiple antihypertensive medications and DM necessitating insulin injections several times daily.

S. Jahan, *Endocrinology Case Studies*,
https://doi.org/10.1007/978-981-95-2770-0_11

Q1. What are the other important points to be looked for in her history?
  (a) Steroid intake
  (b) Coarsening of skin and seborrhea
  (c) Low trauma fracture and loss of height
  (d) Menstrual irregularity and infertility
  (e) Galactorrhea

The lady had a typical picture of Cushing syndrome including such catabolic features as proximal myopathy and metabolic features like HTN, DM, and centripetal obesity. Other cardinal points include easy bruisability, osteoporosis evidenced by low trauma fracture, or loss of height due to vertebral compression collapse. Hyperandrogenic feature like hirsutism, menstrual irregularity or infertility, and thin skin with wide, purple, depressed striae over the abdomen or thigh indicative of protein breakdown are also suggestive of Cushing syndrome. On the other hand, galactorrhea occurs in hyperprolactinemia, and coarsened facies with seborrhea are typical of acromegaly and unlikely to occur in Cushing syndrome. Examination revealed a lady with classic Cushingoid habitus with BMI > 27 kg/m^2, BP—200/100 mm Hg, thinned skin, ecchymosis, and wide purple striae (Figs. 11.1, 11.2, 11.3, and 11.4).

After admission, she underwent investigations to establish the diagnosis. She had moderately elevated 24 h urinary free cortisol, nonsuppressed plasma cortisol on

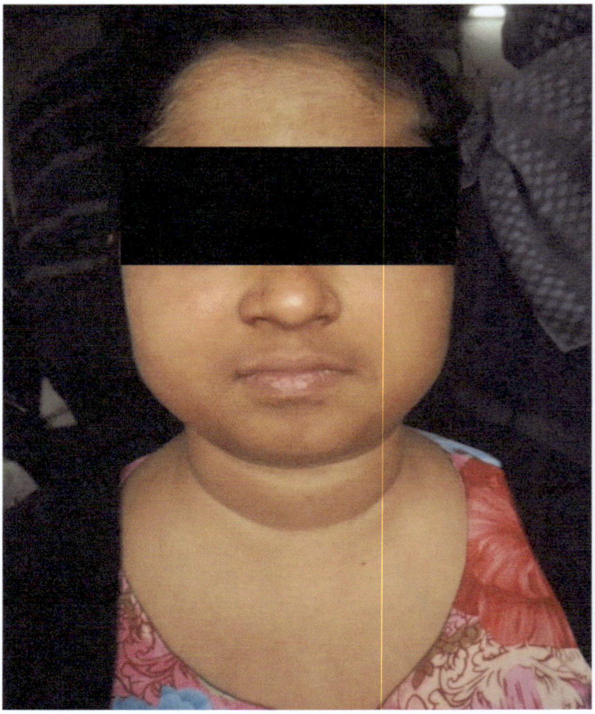

Fig. 11.1 Moon face with supraclavicular fullness

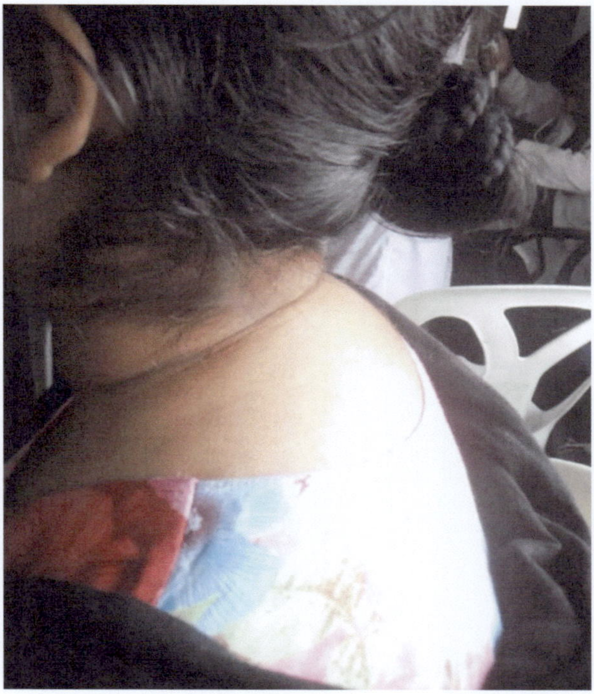

Fig. 11.2 Dorso-cervical fat pad and acanthosis nigricans

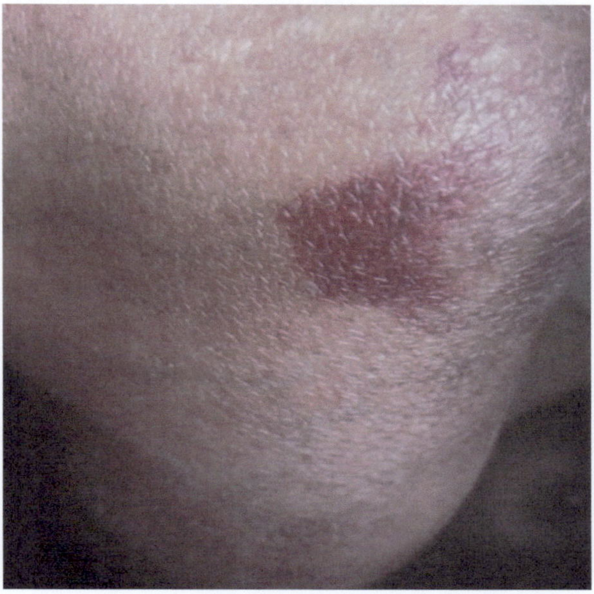

Fig. 11.3 Spontaneous bruise over extremity

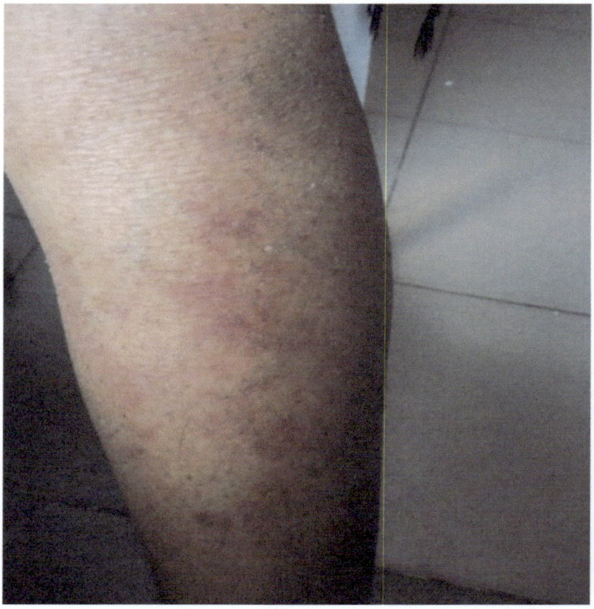

Fig. 11.4 Red purple striae over calf

Table 11.1 Biochemical report of the patient confirming ACTH-dependent Cushing syndrome

Parameter	Patient's value	Reference range
Serum electrolyte	Na—139, **K—2.9**, Cl—95, HCO3—32 mmol/L	Na—135–145, K—3.5–5.1, Cl—95–107, TCO2—22–30 mmol/L
Basal cortisol	**1089 nmol/L**	138–690 nmol/L
Serum cortisol on LDDST	**250 nmol/L**	Nonsuppressed: >50 nmol/L
Evening serum cortisol	**567 nmol/L**	–
24 h UFC	**299 mcg/day**	50–190 μg/ 24 h
Plasma ACTH	**92 pg/dL**	0–46 pg/mL
Serum cortisol on HDDST	**566.3 nmol/L**	Nonsuppressed: >50% decremental value
DHEAS	**344 μg/dL**	70–495 μg/ dL

overnight low-dose dexamethasone suppression test, and loss of diurnal variation of cortisol secretion, thereby confirming Cushing syndrome biochemically (Table 11.1).

Q2. What other hormonal abnormalities are found in Cushing syndrome?
   (a) High TSH
   (b) High LH, FSH, and low estradiol
   (c) Exaggerated response to GH stimulation test
   (d) High AVP
   (e) Variable prolactin

High cortisol level suppresses other hormonal axes; thereby, TSH is supposed to be low, causing secondary hypothyroidism; inhibition of the hypothalamic–pituitary–gonadal axis may lead to hypogonadotropic hypogonadism and suboptimal GH response leading to growth arrest in children. It also inhibits AVP secretion from the hypothalamus which often masks DI in the patient with panhypopituitarism. There might be *disconnection hyperprolactinemia* resulting from stalk compression by macroadenoma in Cushing disease or low prolactin due to the compressive effect on surrounding lactotrophs.

Q3. What is the most important investigation to find out the etiology of Cushing syndrome?
   (a) Dynamic pituitary MRI
   (b) Adrenal CT scan
   (c) Plasma ACTH
   (d) High-dose dexamethasone suppression test (HDDST)
   (e) Inferior petrosal sinus sampling (IPSS)

The single most important test to find out the etiology is plasma ACTH differentiating ACTH-dependent from ACTH-independent Cushing syndrome. Other tests mentioned above are mostly used for localization of the source, e.g., adrenal/pituitary imaging. HDDST and IPSS are done to categorize ACTH-dependent Cushing into pituitary or ectopic variety. Her plasma ACTH was 92 ng/dL which favored ACTH-dependent Cushing syndrome. She also had hypokalemic alkalosis on several occasions. Overnight HDDST led to a decrement of plasma cortisol by 52% albeit not <50% from the baseline value (Table 11.1). Dynamic pituitary MRI revealed a 5 mm microadenoma (Fig. 11.5), and interestingly her adrenal CT scan also showed an adrenal nodule on the right side (Fig. 11.6).

Q4. What further clinical clues should you look for based on the above results?
   (a) Palpable lump in abdomen
   (b) Hyperpigmentation
   (c) Fluctuating HTN
   (d) Cough, hemoptysis, and chest pain
   (e) Clitoromegaly

Ectopic Cushing syndrome is a rare phenomenon classically occurring in the context of small-cell carcinoma of lungs, bronchial/thymic carcinoid, and other neuroendocrine tumors. Short duration and rapid progression of the disease are likely if not due to a benign entity, e.g., bronchial carcinoid. In contrast to Cushing disease that is more common in a young female, it tends to occur in older males having a history of smoking. Weight loss with severe proximal myopathy and striking hyperpigmentation along with such localizing features of primary tumor- like cough, hemoptysis, or chest pain suggest the underlying diagnosis, e.g., oat cell carcinoma of the lung. On the other hand, the presence of nonspecific headache, hirsutism, or menstrual abnormalities in a young

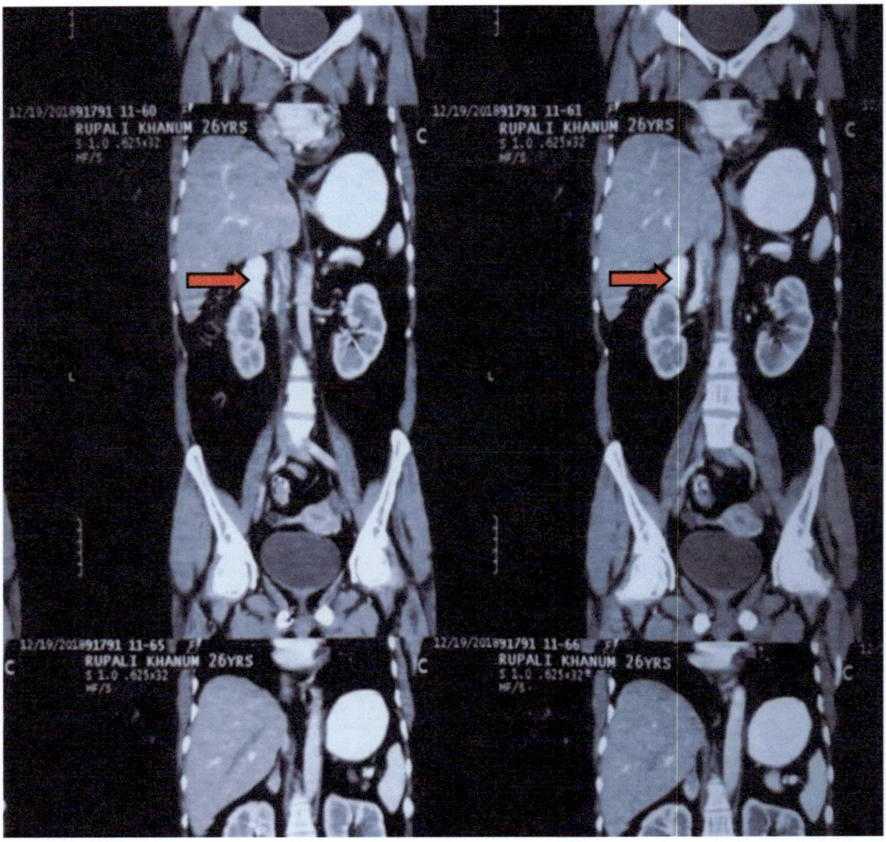

Fig. 11.5 Right adrenal nodule on CT of abdomen

female point toward Cushing disease where the course is relatively indolent. An abdominal lump indicates adrenocortical carcinoma in such a patient that might lead to clitoromegaly and other signs of virilization. However, high ACTH in this case virtually excluded such diagnosis, making ACTH-independent Cushing a rare possibility. Fluctuating HTN is the hallmark of pheochromocytoma, which was not a possibility in this lady as adrenocortical and medullary tumors are unlikely to coexist.

Considering all the available reports, Cushing disease was our final diagnosis. It should be mentioned that in 10% of CD, hypokalemia can occur and plasma cortisol may not drop to the desired value (<50% from baseline) on HDDST. However, the presence of adrenal nodules created some confusion.

Q5. What other investigations might prove helpful?
  (a) Adrenal venous sampling (AVS)
  (b) NP-59 scan
  (c) MIBG scan

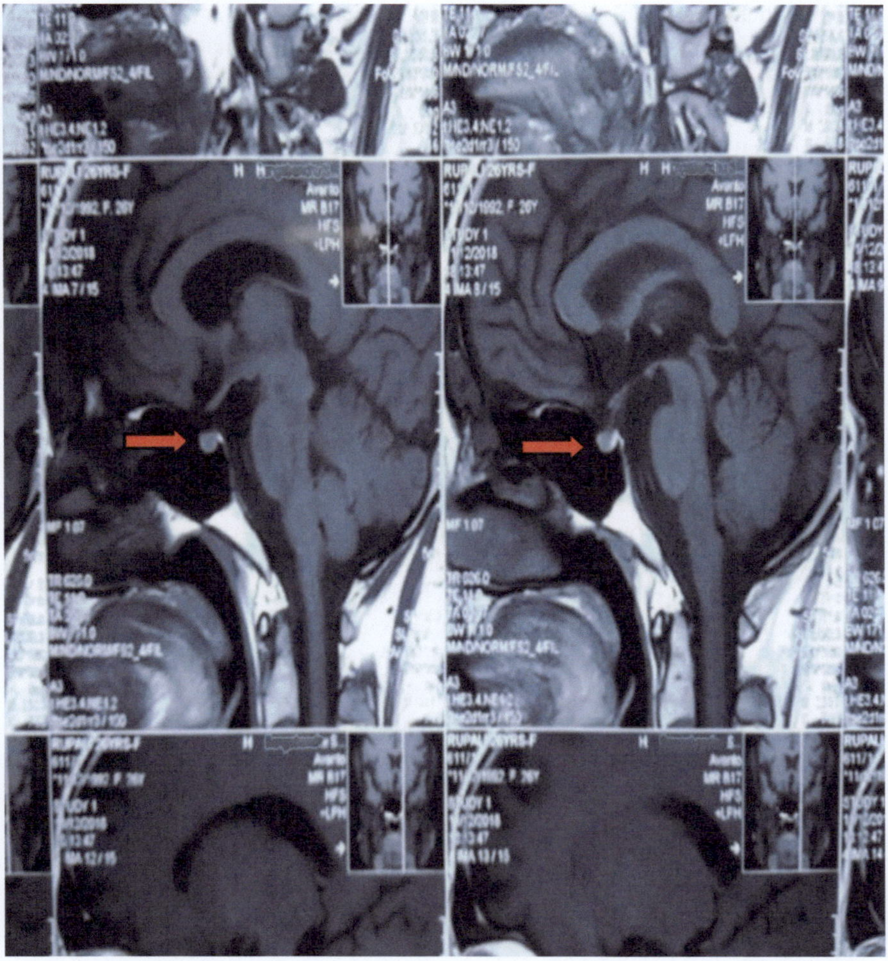

Fig. 11.6 Small pituitary tumor on dynamic MRI

(d) Immunostaining of resected tumor
(e) Serial follow-up of hormone and adrenal nodule

The imaging phenotype of this adrenal nodule suggested a lipid-poor adenoma {(2.6 × 2.6 × 2.1) cm, well-defined margin, moderately heterogeneous, HU+ 40, washout—100%}. AVS is sometimes performed in difficult cases of ACTH-dependent Cushing syndrome like: bilateral macronodular adrenal hyperplasia (BIMAH) where there is bilateral adrenal nodule and primary pigmented nodular adrenal disease (PPNAD) where the adrenals are normal on CT or in an adrenal incidentaloma presenting with subclinical Cushing syndrome to localize the side of autonomy. Cortisol lateralization ratio >2.3 would indicate unilateral hypersecretion on AVS in such cases. NP-59 scintigraphy employs ^{113}16, β-nor-iodocholesterol molecule that is helpful in showing radiotracer accumulation in the hyperfunctioning nodule. It can be done in cases of bilateral adrenal incidentaloma, e.g., bilateral adrenal metastasis or identifying laterality of a hypersecretory nodule. In our case, both the tests were not necessary as we were dealing with ACTH-dependent rather than ACTH-independent/adrenal Cushing. Ultimately, the conclusion was made from postsurgical immunophenotyping of the pituitary microadenoma that stained positive for ACTH and showed Crooke's hyaline change. During postoperative follow-up, there was significant clinical and biochemical improvement of Cushing syndrome and regression of adrenal nodule. There is evidence that long-standing ACTH hypersecretion might lead to diffuse followed by nodular hyperplasia of the adrenals with the ultimate development of autonomous nodules.

Discussion

When iatrogenic causes are excluded, the most common cause of Cushing syndrome is *Cushing disease*, accounting for approximately 70% of the cases [1]. Most tumors are small ACTH-secreting microadenomas (<1 cm); however, in 10% cases, the tumors can be large macroadenomas which are more invasive in nature. Selective surgical removal of a microadenoma results in cure with a very low recurrence rate [2]. The disease typically occurs in a young female presenting with centripetal obesity, HTN, glucose intolerance, easy bruising, osteoporosis, or fracture. The features specific for CD are related to adrenal hyperandrogenism like hirsutism, oligomenorrhea, or infertility. Briefly, the diagnosis of an ACTH-secreting pituitary tumor is suggested by features of hypercortisolism, elevated 24 h urinary free cortisol levels, and/or elevated late-night salivary cortisol values together with nonsuppressed serum ACTH levels. Failure to suppress the morning cortisol level to less than 50 nmol/L after administration of 1 mg dexamethasone at 11 pm supports the diagnosis [1].

The adrenal glands are normal or diffusely hyperplastic in CD resulting from chronic ACTH stimulation. Sometimes, prolonged stimulation may lead to the formation of single or multiple nodules in a background of hyperplasia. This occurs in 10–40% of patients with Cushing disease, and the nodule may be up to several centimeters in diameter [3–6]. Patients tend to be older and to have had symptoms for a longer time, but they otherwise present with the classic clinical features of Cushing syndrome. This contrasts with the *adrenal nodule* in ACTH-independent Cushing syndrome where the contralateral and rest of the ipsilateral glands are atrophic. Adrenal senescence is associated with increasing prevalence of incidentaloma which when exposed to chronically high ACTH may turn into macroadenoma that even sometimes may become autonomous. These nodules are more common in older age groups having long-standing Cushing syndrome, and HTN and alopecia are supposed to be more common in them. The difficulty in diagnosing the nature of the nodule is due to their nonsuppressibility on HDDST, and ACTH may be normal or even suppressed. However, they arise because of sustained high corticotroph activity, and treatment should be pituitary adenomectomy rather than adrenalectomy.

Answers

Q1. (a), (c), (d)
Q2. (c)
Q3. (c)
Q4. (b)
Q5. (d)

References

1. Newell-Price J, Bertagna X, Grossman AB, et al. Cushing's syndrome. Lancet. 2006;367:1605–17.
2. Woo YS, Isidori AM, Wat WZ, et al. Clinical and biochemical characteristics of adrenocorticotropin-secreting macroadenomas. J Clin Endocrinol Metab. 2005;90:4963–9.
3. Aron DC, Findling JW, Fitzgerald PA, et al. Pituitary ACTH dependency of nodular adrenal hyperplasia in Cushing's syndrome: report of two cases and review of the literature. Am J Med. 1981;71:302–6.
4. Doppman JL, Nieman LK, Travis WD, et al. CT and MR imaging of massive macronodular adrenocortical disease: a rare cause of autonomous primary adrenal hypercortisolism. J Comput Assist Tomogr. 1991;15:773–9.
5. Samuels MH, Loriaux DL. Cushing's syndrome and the nodular adrenal gland. Endocrinol Metab Clin North Am. 1994;23:555–569. 225.
6. Sturrock ND, Morgan L, Jeffcoate WJ. Autonomous nodular hyperplasia of the adrenal cortex: tertiary hypercortisolism? Clin Endocrinol. 1995;43:753–8.

12 A 46-Year-Old Male with Paroxysmal Abdominal Pain, Headache, Sweating, and Palpitations

Introduction

This is an interesting case of a middle-aged shopkeeper presented with a recurrent panic attack. Clinically, he was normotensive without evidence of any HTN-related target organ damage. However, the panic attacks were very much suggestive of typical "Pheo spell" having a high positive predictive value for pheochromocytoma. Though his urinary catecholamine was minimally elevated, the imaging phenotype highly favored this diagnosis. Initially, operation was denied by the surgeons doubting a normotensive Pheo; later on, he was operated based on the concerning imaging characters. The diagnosis was confirmed by sudden hypertensive emergency during the procedure that was promptly relieved following the removal of the tumor having typical macroscopic as well as microscopic appearance of pheochromocytoma.

Case Scenario

A 46-year-old shopkeeper presented with recurrent abdominal pain and vomiting for the last 1 month. He complained of episodic upper abdominal pain, burning in nature unrelated to food. However, it was induced by exercise and stress and tend to remit with rest. Each episode was associated with throbbing headache with palpitations and subsided with profuse sweating. The spells would occur two to three times a day, and each one persisted for 10–30 min. During several episodes, he experienced fear of impending death.

On examination, his BMI was 22.6 kg/m^2, pulse—96/min, regular and high volume in nature, BP—125/80 mm of Hg (supine)/140/70 mm of Hg (standing). There was no thyromegaly, proptosis, or tremor. His systemic examination including fundoscopy was unremarkable.

S. Jahan, *Endocrinology Case Studies*,
https://doi.org/10.1007/978-981-95-2770-0_12

Q1. What is the probable diagnosis?
   (a) Thyrotoxicosis
   (b) Peptic ulcer disease
   (c) Pheochromocytoma
   (d) Panic attack
   (e) Episodic migraine

Thyrotoxicosis was unlikely provided; there was no thyromegaly, orbitopathy, or any features of toxicosis other than sweating and palpitations occurring in episodes. The attacks were not precipitated by any psychological or situational cues, making panic attacks less likely. Though repeated abdominal pain and vomiting could be attributed to peptic ulcer disease, other symptoms were unlikely to occur because of this. Considering the characteristics of the episodes, they probably represented classic "Pheo" spells. Such spells are highly sensitive and specific for Pheos (>95%). However, his ambulatory BP monitoring failed to reveal any hypertensive episodes.

Q2. Which of the following statement regarding pheochromocytoma is true?
   (a) Fluctuating HTN is the most common in pheochromocytoma.
   (b) Persistent HTN is the most common in pheochromocytoma.
   (c) Normotensive pheochromocytoma is the commonest category.
   (d) Normotensive pheochromocytoma occurs rarely.
   (e) Hypotension virtually excludes pheochromocytoma.

Contrary to the popular belief, fluctuating HTN occurs in only 30% of pheochromocytoma, whereas in 50% of cases, HTN is persistent. In about 20% of patients, BP is normal on ambulatory monitoring, and the performance of HTN as a screening criterion is poor. Sometimes, these patients can also present with hypotension owing to volume depletion resulting from long-standing vasoconstriction, post-ictal natriuresis, altered vasomotor reflexes, or the dopamine-secreting nature of the tumor.

His routine investigations including serum electrolytes were normal. He had a USG of the abdomen done previously showing a right adrenal incidentaloma. Considering the high pretest probability of pheochromocytoma, we moved on to hormonal testing. His thyroid function test was normal. Basal serum cortisol, plasma ACTH, and DHEA-S reports were also unremarkable. However, his 24 h urinary metanephrine was at the upper limit of the reference range. ECG showed ST depression and T wave inversion in leads I, II, III, and V5–V6 (Table 12.1).

Q3. What is the ideal biochemical screening test for pheochromocytoma?
   (a) Plasma metanephrine and nor-metanephrine
   (b) 24 h urinary metanephrine and nor-metanephrine
   (c) 24 h urinary VMA
   (d) 24 h urinary catecholamine
   (e) Plasma catecholamine

Table 12.1 Biochemical report of the patient showing minimally high urinary catecholamine

Parameter	Patient's value	Reference range
Serum electrolyte	Na—139, K—4.2, Cl—104, CO_2—28 mmol/L	Na—135–145, K—3.5–5.1, Cl—95–107, TCO_2—22–30 mmol/L
FT4	1.22 ng/dL	0.8–1.8 ng/dL
TSH	0.89 µiu/mL	0.35–5.5 µIU/mL
Basal cortisol	362 nmol/L	138–690 nmol/L
Plasma ACTH	32 pg/mL	0–46 pg/mL
DHEAS	99 ug/dL	106–464 ug/dL
24 h urinary metanephrine	**1737.07 nmol /day**	Normal <1775 nmol/day
24 h urinary nor-metanephrine	225 ug/day	75–375 ug/day

The ideal screening test for pheochromocytoma is 24 h urinary metanephrine having high sensitivity and specificity. On the other hand, plasma metanephrine has relatively low specificity (though high sensitivity) giving rise to more false-positive results, increasing the cost of unnecessary investigations. Urinary VMA is no longer routinely recommended because of its poor sensitivity and specificity. In our patient, though the urinary metanephrine was not too high, we proceeded with adrenal CT considering the high pretest probability.

Q4. What could be the explanation behind minimally elevated plasma/urinary metanephrine in pheochromocytoma?
   (a) Erroneous assay technique
   (b) Intratumoral catecholamine metabolism
   (c) Nonsecretory pheochromocytoma
   (d) Episodic secretion of catecholamine
   (e) All the above

Various drugs including psychoactive medications, physical/emotional stress, and surgery alter the biochemical results for catecholamine or its metabolite. The newer assay method of HPLC/TMS is less prone to bias to drugs and dietary tyramine, and biochemical testing is possible while the patient is getting them. Active hormones often undergo intratumoral metabolism in a large tumor pouring only inactive metabolites into circulation not identified by commercial assays. Paraganglioma in the head and neck region is often nonsecretory, presenting only with compressive symptoms. On the other hand, malignant/poorly differentiated tumors often lack enzymatic machinery for hormonal activation. If the sample is not collected during a spell, then it might also miss the diagnosis. So, in case of a high index of suspicion, adrenal imaging is warranted anyway. In our patient, it showed a well-defined iso- to hypodense mass (about 4.4 cm × 3.9 cm) in the right suprarenal region; precontrast HU was +30 with heterogeneous contrast enhancement and nonenhancing area of necrosis, and contrast washout was 44%, suggestive of pheochromocytoma (Fig. 12.1).

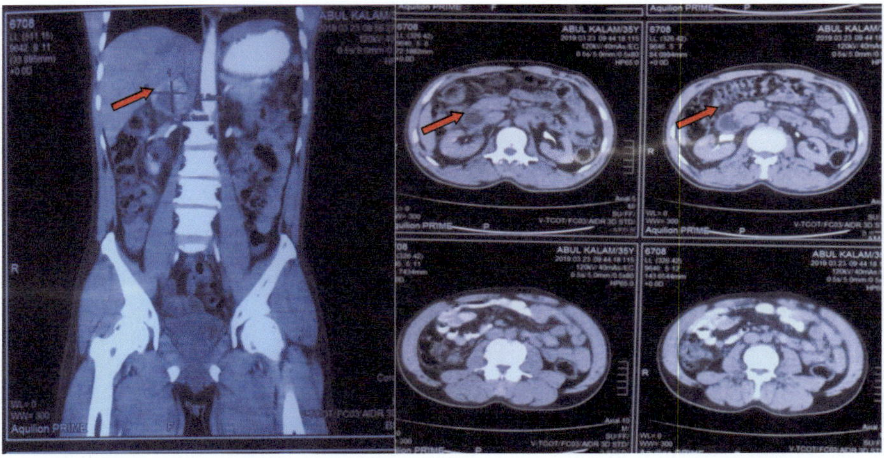

Fig. 12.1 Right adrenal mass suggestive of pheochromocytoma

Q5. The probable feature of malignancy in this patient could be predicted from:
   (a) Size >4 cm
   (b) Heterogenous nature with areas of necrosis
   (c) Contrast washout <50%
   (d) Precontrast HU >30
   (e) Absence of HTN

This pheochromocytoma was likely to be a malignant one considering the size, heterogeneous nature, and delayed contrast washout. The absence of HTN often points to the primitive, undifferentiated nature of the tumor lacking dopamine β-hydroxylase enzyme or intratumoral metabolism of the active hormone. However, the density of Pheos, irrespective of its benign/malignant behavior, tends to be high >30 HU, making it a poor marker of malignancy. T2-weighted MRI shows bright hyperintensity, the so-called "light bulb sign" that characterizes pheochromocytomas in doubtful cases of CT scan. The patient was referred to an adrenal surgeon and was initially denied surgery as the surgeon was doubtful of a normotensive pheochromocytoma with minimally elevated catecholamine. However, the size and other imaging characters were alarming enough to demand surgical removal.

So, finally, he underwent surgery with the removal of a typical pheochromocytoma. During the procedure, his BP shot to 260/160 mm Hg, and it was managed with I/V labetalol and GTN, and BP came down to 120/80 mm Hg in the immediate postoperative period. The resected tumor was friable with an angry red appearance. The ultimate diagnosis was made retrospectively by typical histopathologic features. Immunohistochemical markers were also positive as expected. He is normotensive and symptom-free till now (Figs. 12.2, 12.3, and 12.4).

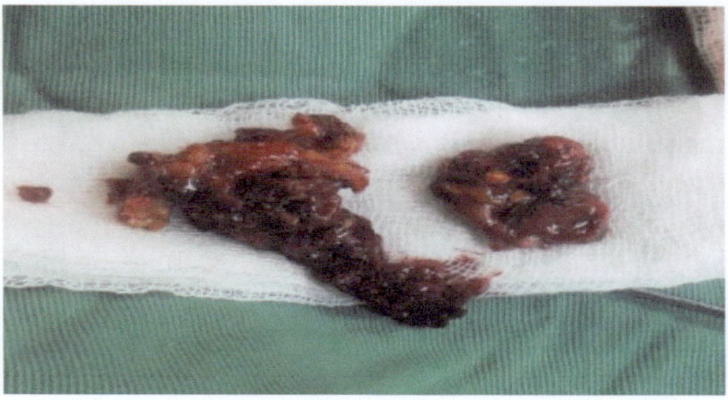

Fig. 12.2 The resected tumor fragile in nature and dark brown in color

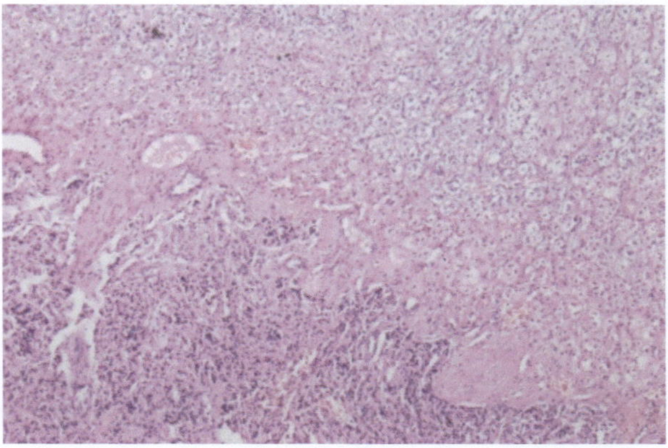

Fig. 12.3 Pheochromocytoma abutting unremarkable adrenocortical tissue

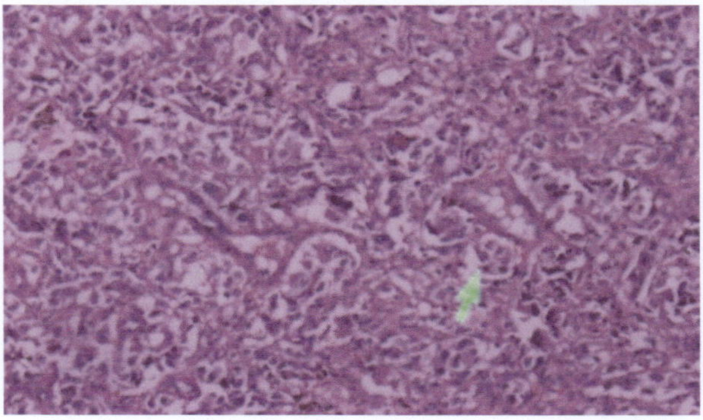

Fig. 12.4 Tumor with typical Zelballen appearance and fibrovascular stroma

Discussion

Catecholamine-secreting tumors that arise from the chromaffin cells of the adrenal medulla and the sympathetic ganglia are referred to as pheochromocytomas and catecholamine-secreting paragangliomas, respectively [1]. They are rare, with an annual incidence of 2 to 8 cases per one million people [2]. Based on screening studies for the secondary causes of hypertension in outpatients, the prevalence of pheochromocytoma has been estimated at 0.1% to 0.6%. These tumors occur with equal frequency in men and women, primarily in the third, fourth, and fifth decades [3]. In the Mayo Clinic experience, among 54 patients with pheochromocytoma on whom autopsy was performed, the tumor was not suspected in 75% of patients and contributed to death in 55% of patients [4]. Contrary to the popular belief, not all pheochromocytomas have paroxysmal HTN; they can have persistent HTN or may be normotensive or hypotensive as well. The lability in blood pressure can be attributed to the episodic release of catecholamines, chronic volume depletion, and impaired sympathetic reflexes. In addition to volume depletion, altered sympathetic vascular regulation may have a role in orthostasis, which is frequently observed in patients with pheochromocytoma [5]. Pheochromocytoma may also produce several vasodilator compounds (e.g., dopamine, calcitonin gene-related peptide, vasoactive intestinal peptide, adrenomedullin, met-enkephalin, atrial natriuretic peptide, and neuropeptide Y) that can influence vascular responsiveness [6, 7]. Angiotensin II perhaps also plays a contributory role by enhancing the release of norepinephrine at presynaptic sites [8]. Desensitization of the cardiovascular system to a persistently high level of circulating catecholamine may also contribute to the absence of HTN [9]. The *normotensive Pheo* is diagnosed based on the classic "Pheo" spell, elevated plasma/urinary catecholamine and their metabolite, and the typical imaging phenotype. The biochemical report might be normal as well due to various reasons like assay error, episodic hormonal surge, intratumoral metabolism, etc. In such a case, emphasis should be given to other suggestive clinical features and imaging characteristics. The preoperative management of normotensive Pheo is still a matter of debate; some authorities are in favor of α-blockade, while others oppose it as it might lead to more hypotensive episodes postoperatively.

Answers

Q1. (c)
Q2. (b)
Q3. (b)
Q4. (c)
Q5. (a), (b), (c), (e)

References

1. Lloyd RV, Tischer AS, Kimura N, et al. Adrenal tumors: introduction. In: DeLellis RA, Lloyd RV, Heitz PU, et al., editors. World Health Organization classification of tumours: pathology and genetics of tumours of endocrine organs. Lyon: IARC Press; 2004. p. 136–8.
2. Stenstrom G, Svardsudd K. Phaechromocytoma in Sweden, 1958-1981: an analysis of the National Cancer Registry data. Acta Med Scand. 1986;220:225–32.
3. Sinclair AM, Isles CG, Brown I, et al. Secondary hypertension in a blood pressure clinic. Arch Intern Med. 1987;147:1289–93.
4. Sutton MG, Sheps SG, Lie JT. Prevalence of clinically unsuspected pheochromocytoma: a review of a 50-year autopsy series. Mayo Clin Proc. 1981;56:354–8.
5. Munakata M, Aihara A, Imai Y, et al. Altered sympathetic and vagal modulations of the cardiovascular system in patients with pheochromocytoma: their relations to orthostatic hypotension. Am J Hypertens. 1999;12:572–80.
6. Goldstein DS, Stull R, Eisenhofer G, et al. Plasma 3-4 dihydroxyphenylalanine (DOPA) and catecholamines in neuroblastoma and pheochromocytoma. Ann Intern Med. 1986;105:887–8.
7. Lundberg JM, Hokfelt T, Hemsen A, et al. Neuropeptide Y-like immunoreactivity in adrenaline cells of adrenal medulla and in tumors and plasma of pheochromocytoma patients. Regul Pept. 1986;13:169–82.
8. Bravo EL, Tarazi RC, Fouad FM, et al. Blood pressure regulation in pheochromocytoma. Hypertension. 1982;4(suppl II):193–9.
9. Tsujimoto G, Honda K, Hoffman BB, et al. Desensitization of postjunctional a1- and a2-adrenergic receptor mediated vasopressor responses in rats harboring pheochromocytoma. Circ Res. 1987;61:86–98.

A 48-Year-Old-Male Presented with High Serum Calcitonin Level 13

Introduction

This case highlights the role of calcitonin in the follow-up and management of thyroid medullary carcinoma. The progressive rise in the serum calcitonin level during the postsurgical follow-up in a 48-year-old male led to his hospitalization, and procedures like CT scan and isotope bone scan established MTC recurrence with metastasis. He was given palliative care with tyrosine kinase inhibitors. There is elaborate discussion on the nature of thyroid malignancies, presentation and management of MTC, and the conditions associated with the high calcitonin level.

Case Scenario

A 48-year-old schoolteacher was diagnosed with medullary thyroid carcinoma (MTC) 12 years back treated by total thyroidectomy. He stated that he had a rapidly growing thyroid nodule without any evidence of thyroid dysfunction at that time, which proved to be MTC after surgery.

Since then, he had been on regular follow-up and remained symptom-free. He was on a regular replacement dose of levothyroxine for postsurgical hypothyroidism and was diabetic for the last 2 years on irregular medication.

Q1. Which of the following statements regarding MTC is true?
- (a) It arises from the Oxyphil cell
- (b) Lymphatic spread is more common than hematogenous spread
- (c) Tends to occur in the upper two-thirds of the thyroid gland
- (d) It has the avidity for radio-iodine isotope
- (e) Commonly occurs as part of a hereditary syndrome

S. Jahan, *Endocrinology Case Studies*,
https://doi.org/10.1007/978-981-95-2770-0_13

MTC is a distinct variety of thyroid malignancy that arises from parafollicular C cells of the thyroid rather than the oxyphil cells of the parathyroid. In comparison to thyroid follicular carcinoma, it has a very aggressive behavior with early lymphatic as well as hematogenous spread. It follows the anatomic location of C cells, thereby arising commonly in the upper two-thirds of the gland. Only 20% occur as part of hereditary disorders like MEN 2A/2B, whereas the rest are sporadic. They are distinct from thyroid follicular carcinoma because they do not trap radioiodine. So, radioiodine is useful neither in their diagnosis nor their management.

His diagnosis was confirmed following the histopathology of the resected tumor.

Q2. What are the histopathologic characteristics of MTC?
- (a) Presence of psammoma bodies
- (b) Presence of spindle cells with amyloid stroma
- (c) Presence of spindle cells with ropy colloid
- (d) Positive immunostaining for calcitonin
- (e) Presence of overlapping nuclei with pseudo-inclusion bodies

MTC belongs to a category of neuroendocrine tumor (NET) that is characterized by abundant amyloid stroma with scattered spindle cells (Fig. 13.1). On the other hand, papillary thyroid carcinoma is a type of differentiated thyroid cancer (DTC) that has large overlapping nuclei with powdery chromatin often having an "Orphan Annie eye" appearance. It may also have a groove in the nuclear membrane producing pseudo-inclusion bodies and laminated, concentric calcium deposits in the cytoplasm called "psammoma body" (Fig. 13.2). The definitive character of MTC is positive immunostaining for calcitonin that is a good biomarker for diagnosis and follow-up.

Serum calcitonin measured during the annual follow-up was normal for the initial 2 years after surgery. However, in the following years, its level started to get

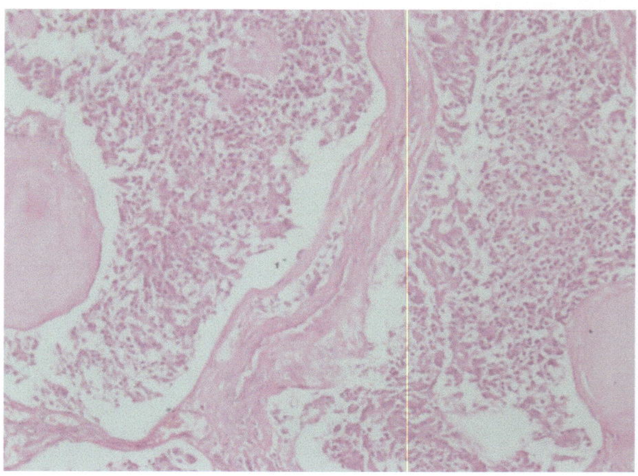

Fig. 13.1 MTC showing amyloid stroma with spindle cells

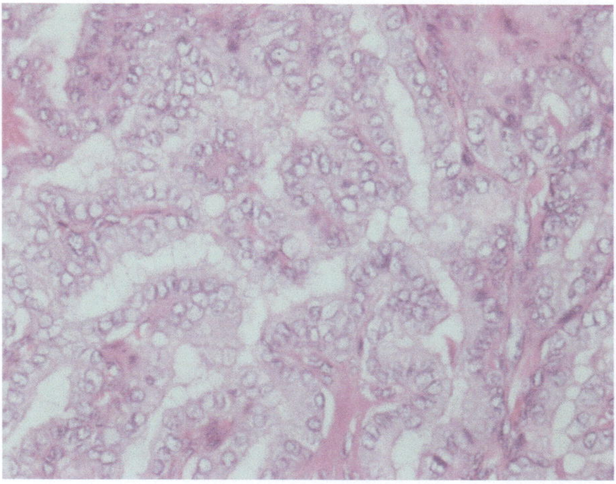

Fig. 13.2 Nuclear features of papillary thyroid carcinoma with crowded overlapping nuclei, nuclear grooves (arrow), pseudo-inclusions and vacuoles

elevated, and recently it became much higher for which he was asked to take admission for further evaluation. On query, he denied having any fever, weight loss, bone pain, jaundice, cough, chest pain, or hemoptysis. None of his family members had any similar type of illness.

Q3. What other clues should be looked for in his history?
- (a) Watery diarrhea
- (b) Painful thyroid nodule
- (c) Painless thyroid nodule
- (d) Flushing
- (e) Bloody diarrhea

MTC is a neuroendocrine tumor that secretes a plethora of peptide hormones like serotonin, bradykinin, tachykinin VIP, ACTH, gastrin, etc. These hormones might lead to various systemic manifestations like wheezing, flushing, or watery diarrhea due to serotonin or VIP secretion. The patient may present with hyperpigmentation, hypokalemic HTN, osteoporosis, or proximal myopathy resulting from ectopic Cushing syndrome. Unlike the painless thyroid nodules in DTC, the nodule in these cases is often painful. On examination, he was a well cooperative gentleman. There was a thyroidectomy scar mark in front of the neck; no lymphadenopathy or thyromegaly was present. His pulse rate was 88/min, and BP was 120/80 mm Hg; BMI—22.4 kg/m^2. All other systemic examinations revealed no abnormality. He was investigated accordingly after admission. He had hypocalcemia, hypophosphatemia with high-intact PTH, and normal thyroid function; USG of the neck showed few sub-centimeter lymph nodes in the central cervical group (Table 13.1). There was a progressive rise in his serum calcitonin level, and at admission, it was found to be remarkably high (Table 13.2).

Table 13.1 Biochemical report of the patient showing secondary hyperparathyroidism

Parameter	Results	Reference range
Serum calcium	8.7 mg/dL	9–11 mg/dL
Serum phosphate	2.9 mg/dL	3–4.5 mg/dL
Plasma alkaline phosphatase	79 U/L	50–136 U/L
Serum albumin	42 g/L	35–50 g/L
Plasma-intact PTH	124.8 pg/mL	11–67 pg/mL
Serum TSH	1.89 mIU/mL	0.4–4 mIU/mL
X-ray of chest, P/A view	Normal	–
USG of W/A	Hepatomegaly with fatty liver	–
USG of neck	No thyroid parenchyma. Sub-centimeter LN in left levels VI, V and right level IV.	–

Table 13.2 Rising trend of the serum calcitonin level during the postsurgical follow-up

Date	Name of investigation	Results
OCT/2002	Serum calcitonin	46.2 pg/mL (N:16.2—57.2)
DEC/2002	"	35.2 pg/mL
DEC/2003	"	25.4 pg/mL
AUG/2004	"	299 pg/mL
APRIL/2005	"	232 pg/mL
SEPT/2007	"	322 pg/mL
MARCH/2009	"	319 pg/mL
MAY/ 2010	"	447 pg/mL
FEB/2012	Serum calcitonin	322 pg/mL
JUNE/2012	"	247 pg/mL
OCT/2013	"	167 pg/mL
MAY/2015	"	1116.7 pg/mL
MARCH/12	Isotope whole body scan	There is radiotracer concentration in the lateral aspect of lower end of femur and also in tarsal and metatarsal bones. The rest of the skeletal system shows uniform radiotracer concentration
AUG/2015	Isotope bone scan	Widespread uptake of radiotracers in dorsolumbar spine, right femur, and tarsal and metatarsal bones bilaterally
AUG/2015	S. Vitamin D level	13.6 ng/mL

Q4. What are the factors that might contribute to moderately high calcitonin levels in the initial years?
(a) Chronic renal failure
(b) COPD
(c) MEN 1 syndrome
(d) Autoimmune thyroiditis
(e) Pheochromocytoma

Though serum calcitonin is a very good biomarker for diagnosis and postsurgical follow-up of MTC, it is not 100% specific. Mild-to-moderate elevation can occur in other conditions like smoking, chronic renal failure, COPD, or autoimmune thyroiditis. Pheochromocytoma is a neuroendocrine tumor that may co-secrete calcitonin. So, measurement of serum calcitonin is indicated only under certain situations in thyroid nodules including a positive family history of MEN2A/B, a suggestive clinical feature like diarrhea, or wheezing. MEN 1 syndrome comprises a tumor arising from the parathyroid, pituitary, and pancreas, and it does not encompass MTC.

Q5. What factors might lead to very high calcitonin?
(a) Recurrence/persistence of MTC
(b) Bronchial carcinoid
(c) Small-cell carcinoma of the lung
(d) Squamous cell carcinoma of lungs
(e) Hook effect of the assay

Postsurgical follow-up of MTC should be done with periodic USG of neck and serum calcitonin level at 3 and 6 months initially and annually thereafter. A serum calcitonin level <10 pg/mL at 3 months indicates a possible cure, and a level above this indicates the persistence of the disease. However, if the level tends to rise after an initial period of remission, then it indicates recurrence; level >500 mg/mL indicates gross local spread, whereas level >1000 pg/mL indicates widespread distant metastasis. On the other hand, hook effect sometimes occurs during the assay with a very high level of calcitonin, where it is erroneously reported to be low or normal due to supersaturation of the antibody by the very high antigen level. Our patient developed a very high level of calcitonin 10 years after surgery, and there was activity seen in the bone scan suggestive of possible recurrence with skeletal metastasis.

Q6. What other investigations are necessary at this stage?
(a) CT chest and abdomen
(b) I^{131} isotope scan
(c) CEA, neuron-specific enolase, and chromogranin
(d) CEA, thyroglobulin, and thyroperoxidase
(e) Skeletal survey

On query, the patient admitted of having constant aching pain in his back and buttock that got aggravated on movement, coughing, and straining. A skeletal survey followed by a bone scan was done that revealed widespread uptake of radiotracers in the dorsolumbar spine, right femur, and tarsal and metatarsal bones bilaterally. There was a compression collapse of D10, 11, and 12 vertebra and osteolytic lesion in the right femoral neck. CT of chest showed multiple metastases of variable size and shape in both lung fields which appeared as military mottling on CXR. MTC is a typical NET that elaborates an array of biomarkers including oncofetal protein like CEA, neuron-specific enolase, and chromogranin. There are certain tumors which though aggressive lack calcitonin secretion because of dedifferentiation or because they exclusively secrete other peptides like chromogranin in chromograninoma. The dedifferentiated tumor is characterized by aggressive behavior and high CEA level. Thyroglobulin is the marker for the follow-up of DTC which has little value in monitoring MTC. On the other hand, thyroperoxidase (TPO) is the enzyme needed for the oxidation of iodine during thyroid hormone synthesis. Clinically, anti-TPO Ab test is routinely done for the diagnosis of autoimmune thyroiditis. I^{131} isotope scan is used both for therapeutic and diagnostic purposes in DTC like papillary/follicular carcinoma arising from a thyroid follicular cell that avidly uptakes radioiodine. It has got no utility in the management of C-cell tumors like MTC.

Q7. How should he be treated now?
- (a) Using suppressive dose of levothyroxine
- (b) Tyrosine kinase inhibitor: vandetanib
- (c) Palliative therapy for bone pain
- (d) Chemoembolization
- (e) Neck dissection

This patient presented with widespread metastatic malignancy where any curative therapy was deemed impossible. Increasing the dose of LT4 would not be helpful as the growth of MTC is not TSH-dependent. Chemoembolization could have been attempted if there was a relatively large solitary metastasis rather than a diffuse spread. Similarly, the neck lymph node was subcentemetric without any residual thyroid tissue. Thereby, neck dissection would probably not be helpful in the context of disseminated disease elsewhere. Finally, he was offered palliative therapy for bone pain with a strong analgesic, and tyrosine kinase inhibitor therapy was initiated which he later discontinued because of financial constraints. He succumbed to death a few months later while on follow-up.

Discussion

The thyroid gland comprises mainly of two types of cells, namely the follicular epithelial cells and the parafollicular C cells. The C cells arise from the neural crest and ultimobranchial body and make up 1% of the thyroid cell population.

They secrete a polypeptide hormone consisting of 32 amino acids named calcitonin. *Calcitonin* lacks any significant physiologic role in humans except in pregnancy where it is implicated in strengthening the maternal and fetal skeleton. The neoplastic proliferation of C cells leads to *medullary thyroid carcinoma (MTC),* contributing to 1–3% of total thyroid neoplasia [1]. MTC can be caused by an activating mutation of the RET proto-oncogene, embracing a familial form in 25% of cases, either isolated or in the context of MEN 2 syndromes. The majority (75%) of MTC appears, however, sporadically, with no mutation described [1]. MTC is a prototype neuroendocrine tumor that can secrete a multitude of peptide hormones like serotonin, VIP, gastrin, insulin, catecholamine, ACTH, etc. So, the presentation is often varied and multisystemic including ectopic ACTH syndrome, secretory diarrhea, wheezing, flushing, spontaneous hypoglycemia, fluctuating HTN, etc. Calcitonin is a useful biomarker for the detection and follow-up of MTC, having an important prognostic implication as well. When CT reaches values above 100–200 pg/mL, the presence of MTC is usually certified. CT levels generally parallel with the tumor volume, as well as the appearance and amplitude of metastases [1, 2]. Very rarely, other non-C-cell-derived tumors, such as carcinoids or small-cell lung cancers, may also be accompanied by very elevated CT levels [1]. Lower CT values (between 10 and 40 pg/mL), although considered pathological, are more difficult to be interpreted because the normal range varies depending on sex and the commercial kit used, and slightly increased CT levels may be found in other nonmalignant conditions (certain cases of autoimmune thyroiditis, chronic renal failure, chronic pulmonary disease, acute trauma, pregnancy, hypercalcemia, hypergastrinemia, hyper-catecholaminemia, type 1A pseudohypoparathyroidism, mastocytosis, and sepsis) [1, 3–6]. CT is not the only sensitive marker for MTC, since this type of cancer can also secrete carcinoembryonic antigen (CEA), neuron-specific enolase (NSE), chromogranin A (CgA), and procalcitonin (proCT), all useful for MTC diagnosis and follow-up [5, 6]. Low/normal CT 3 months after surgery suggests cure, and its normalization after only 1 week certifies the absence of lymph node metastasis [1, 2]. Follow-up should be continued every 6 months for the first year and then yearly in cases with undetectable or normal postoperative CT, eventually with a stimulation test for detectable CT $<$ 10 pg/mL. Locoregional recurrence may benefit from surgical reintervention and external radiotherapy [2]. Distant metastases are usually multifocal, but they may progress slowly. Cytotoxic chemotherapy has limited efficacy. Chemoembolization with adriamycin may be efficient in large liver metastases. Kinase inhibitors of tumor cells (RET and other kinases) or of endothelial cells (VEGFR) such as cabozantinib or vadentanib may be efficient in selected cases of progressive disease and should be used as first-line intention [7, 8].

Answers

Q1. (b), (c)
Q2. (b), (d)
Q3. (a), (b), (d), (e)
Q4. (a), (b), (d), (e)
Q5. (a), (c), (e)
Q6. (b), (c)

References

1. Costante G, Meringolo D. Calcitonin as a biomarker of C cell disease: recent achievements and current challenges. Endocrine. 2020;67(2):273–80.
2. Machens A, Lorenz K, Dralle H. Prediction of biochemical cure in patients with medullary thyroid cancer. Br J Surg. 2020; https://doi.org/10.1002/bjs.11444.
3. Kovacs CS, Ralston SH. Presentation and management of osteoporosis presenting in association with pregnancy or lactation. Osteoporos Int. 2015;26(9):2223–41.
4. Verbeek HH, de Groot JWB, Sluiter WJ, Muller Kobold AC, van den Heuvel ER, Plukker JT, Links TP. Calcitonin testing for detection of medullary thyroid cancer in people with thyroid nodules. Cochrane Database Syst Rev. 2020;3:CD010159. https://doi.org/10.1002/14651858.CD010159.
5. Samà MT, Rossetto Giaccherino R, Gallo M, Felicetti F, Maletta F, Bonelli N, Piovesan A, Palestini N, Ghigo E, Arvat E. Clinical challenges with calcitonin-negative medullary thyroid carcinoma. J Cancer Res Clin Oncol. 2016;142(9):2023–9.
6. Gambardella C, Offi C, Patrone R, Clarizia G, Mauriello C, Tartaglia E, Di Capua F, Di Martino S, Romano RM, Fiore L, Conzo A, Conzo G, Docimo G. Calcitonin negative Medullary Thyroid Carcinoma: a challenging diagnosis or a medical dilemma? BMC Endocr Disord. 2019;19(Suppl 1):45.
7. Jayarangaiah A, Sidhu G, Brown J, Barrett-Campbell O, Bahtiyar G, Youssef I, Arora S, Skwiersky S, McFarlane SI. Therapeutic options for advanced thyroid cancer. Int J Clin Endocrinol Metab. 2019;5(1):26–34.
8. Baetu M, Dobrescu R. Novel markers for early diagnosis and prognostic classification in medullary thyroid carcinoma. Acta Endocrinol (Buc). 2017;13(4):519–22.

14 A 47-Year-Old Male with HTN Presented with Unexplained Proteinuria

Introduction

A middle-aged male with relatively well-controlled HTN and DM was evaluated for unexplained microalbuminuria and renal calculi. On scrutiny of his previous reports, we discovered an adrenal incidentaloma and hypokalemia that had escaped attention before. The patient was investigated accordingly and finally found to have an aldosterone-producing adenoma (APA). He underwent unilateral adrenalectomy for APA that led to the resolution of hypokalemic HTN. This case highlights the significance of suspecting primary hyperaldosteronism in a patient with HTN with unexplained target organ damage, e.g., proteinuria.

Case Scenario

A 47-year-old businessman presented with rather difficult-to-control HTN for the last 8 years. He was on ARB and ACE inhibitor and had a positive family history of early-onset HTN. He also had uncomplicated and well-controlled DM, mild prostatic enlargement, and a non-obstructing left renal calculi. However, he had significant proteinuria and was referred to us for evaluation of proteinuria and advice on possible renal biopsy/lithotripsy. His previous investigations showed a normal hemogram, SGPT, serum creatinine, and standard glycemic control (HbA1c: 6.7%). However, his spot UACR was high on at least two occasions: 42.9 and 60.5 mg/mmol, respectively. He had a report of CT scan of abdomen that was done probably to search for any nephropathy, but interestingly it showed a small adrenal nodule. Then, we investigated his serum electrolyte report and found that he had hypokalemia (Na—146, K—2.9, Cl—97, HCO3—37 mmol/L). So, at this point, we dealt with an adrenal incidentaloma that previously escaped attention and evaluation.

S. Jahan, *Endocrinology Case Studies*,
https://doi.org/10.1007/978-981-95-2770-0_14

Q1. What is the probable etiology of this adrenal incidentaloma?
(a) Nonfunctioning adrenal adenoma
(b) Pheochromocytoma
(c) Subclinical Cushing syndrome (SCS)
(d) Aldosterone-producing adenoma (APA)
(e) None of the above

The pretest probability for APA was quite high in this case. He had young-onset HTN with a positive family history of early-onset HTN. HTN was rather uncontrolled with the episode of spontaneous hypokalemia. There was target organ damage, namely microalbuminuria, that remained otherwise unexplained. So, the presence of an adrenal incidentaloma in this setting was highly suggestive of APA, thereby excluding any other possibilities. He was admitted for further evaluation, and ARB/ACE inhibitors were switched to CCB. At the same time, he was given correction for hypokalemia along with liberal salt intake. Then, he underwent three tiers of tests for the diagnosis of APA including screening, followed by confirmatory test and subtype localization.

Q2. Which of the following statements regarding the screening of APA is true?
(a) There is no medication that causes false-positive screening tests.
(b) There is no medication that causes false-negative screening tests.
(c) Calcium channel blockers do not affect the diagnostic accuracy.
(d) Spironolactone must be stopped for at least 6 weeks before testing.
(e) Beta-blockers do not affect diagnostic accuracy.

Though appropriate preparation might be desirable before screening tests, in many cases, it might be attempted without stopping the drugs. Drugs can lead to false-negative results. However, no drug causes a false-positive result. ACE inhibitor and ARB raise PRA or PRC. So, a suppressed value of PRA/PRC in their presence suggests primary aldosteronism. MRAs (e.g. spironolactone and eplerenone) prevent aldosterone from activating the receptor, resulting sequentially in sodium loss, a decrease in plasma volume, and an elevation in renin. If PRA or PRC is not suppressed in a patient treated with an MRA, then no further PA-related testing can be performed, and the MRA should be discontinued for 6 weeks before retesting. However, if the patient is hypokalemic despite the treatment with an MRA, then the mineralocorticoid receptors are not fully blocked, and PRA or PRC should be suppressed in such a patient with PA [1]. Calcium channel blockers and centrally acting α-blockers do not affect the diagnostic accuracy.

The patient had his electrolyte report done repeatedly showing hypokalemia (ranging between 2.1 and 2.9 mmol/L) and an ABG report showing metabolic alkalosis. He had elevated aldosterone with suppressed renin on screening, and later primary aldosteronism was confirmed by nonsuppressed serum aldosterone on the fludrocortisone suppression test (Tables 14.1 and 14.2).

After biochemical confirmation, he underwent an abdominal CT scan in the adrenal protocol. It showed a small left adrenal adenoma (12 × 15 × 13 mm),

Table 14.1 Positive screening test with high aldosterone and low renin

Test	Result	Reference value
Serum aldosterone (pg/mL)	190.9	Early morning—20–180 Upright—30–400
Plasma renin(pg/mL)	<2	4–37.52

Table 14.2 Nonsuppressed plasma aldosterone on fluodrocortisone suppression test

Parameter	Result	Interpretation
Serum aldosterone(pg/mL)	242.30	Suppressed <60

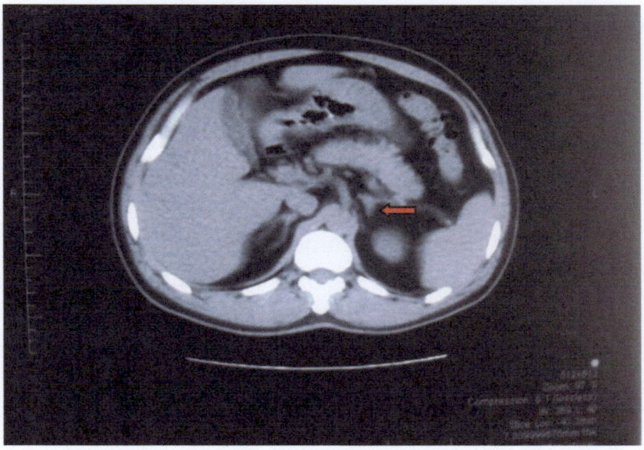

Fig. 14.1 Small left adrenal adenoma

hypodense in character having a HU + 9 to +10 with unremarkable right adrenal (Fig. 14.1). He had a right adrenalectomy performed afterward.

Q3. Which of the following statements is true regarding localization for APA?
In the presence of an adrenal nodule, adrenal venous sampling is not required in:
(a) Young patients (<35 years) with severe hyperaldosteronism
(b) Unilateral, hypodense nodule >1 cm and <2 cm
(c) Unilateral hyperdense nodule >2 cm in size
(d) Prior to the preparation of surgical management
(e) An older patient with a definite nodule

Subtype testing for primary aldosteronism includes an adrenal dedicated abdominal CT scan to exclude adrenocortical carcinoma and to detect APAs in young patients (<35 years of age), followed by AVS when indicated. AVS may not be needed in young patients (<35 years) with marked PA (spontaneous hypokalemia and PAC >30 ng /dL) who have unilateral adrenal macroadenoma (>1 cm and <2 cm) and a normal contralateral adrenal gland. For most patients with PA who want to pursue the surgical option, AVS is indicated to accurately localize the source

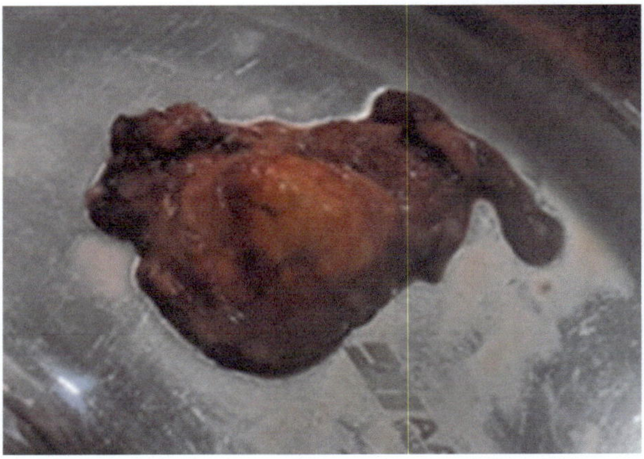

Fig. 14.2 Canary-yellow appearance of the cut surface of the Conn tumor

of aldosterone hypersecretion. The prevalence of adrenal incidentaloma increases with increasing age. So, AVS should be done before surgery in older patients. Our patient had a definite unilateral hypodense nodule with biochemically confirmed PA. So, he proceeded to surgery without undergoing AVS.

The BP of the patient immediately normalized following surgery with the resolution of hypokalemia. On gross examination, the tumor exhibited a typical canary-yellow appearance, and histopathology was suggestive of an adrenal adenoma (Fig. 14.2). At present, he is on regular follow-up without any hypertensive medication and maintaining normokalemia.

Q4. What is the major objective of adrenalectomy in APA?
- (a) Control/remission of HTN
- (b) Remission of hypokalemia
- (c) Prevention of progression of CVS damage
- (d) Averting HTN-related complications
- (e) Improving the quality of life

Normalization of blood pressure should not be the only goal of treatment in APA. In addition to the kidney and colon, mineralocorticoid receptors are present in the heart, brain, and blood vessels. Excessive secretion of aldosterone is associated with an increased risk of cardiovascular disease and morbidity. Therefore, normalization of circulating aldosterone or mineralocorticoid receptor blockade should be part of the management plan for all patients with primary aldosteronism.

Discussion

Primary hyperaldosteronism is characterized by the triad of hypokalemic HTN, high plasma aldosterone, and suppressed renin. However, hypokalemia occurs in only one-third of cases, the rest being normokalemic. Although estimates vary, the prevalence of PA is approximately 5% in patients with hypertension and up to 20% in those with treatment-resistant hypertension [2–4]. Most of the cases are due to aldosterone-producing adenoma (APA) or idiopathic adrenal hyperplasia (IHA), and a small percentage is due to primary adrenal hyperplasia (PAH) or familial hyperaldosteronism. The disease typically presents between 20 and 60 years. There is no reliable clinical phenotype to be looked for. In fact, all hypertensive patients should be screened for this disorder at least once in their life, preferably on the detection of HTN. The diagnosis relies on three sequential tiers of tests, namely screening with plasma aldosterone and renin level/activity ratio, followed by confirmation with plasma/urinary aldosterone nonsuppressibility by various agents, e.g., normal saline or fludrocortisone. Finally, subtype localization is achieved with adrenal CT or adrenal venous sampling when needed. The list of drugs and hormones capable of affecting the renin–angiotensin–aldosterone axis in patients with PA is extensive [5]. However, it is essential for clinicians to understand that although medications used to treat hypertension can potentially cause false-negative testing results in patients with mild PA, there is no medication that causes false-positive results, if a cutoff level for aldosterone is used. Calcium channel blockers and α-1-adrenergic receptor blockers do not affect diagnostic accuracy in most cases. If a solitary small (>1 cm and <2 cm) unilateral nodule with low CT attenuation (<10 HU) on noncontrast CT is found in conjunction with normal contralateral adrenal morphology in a young patient (<35 years) with severe PA [e.g., spontaneous hypokalemia and PAC >832 pmol L − 1 (>30 ng dL − 1)], unilateral adrenalectomy is a reasonable treatment option [6–8]. However, more than 95% of patients with PA are older than 35 years of age. In addition, in young patients with PA, normal-appearing adrenals, minimal unilateral adrenal limb thickening, unilateral microadenomas (≤1 cm), or bilateral macroadenomas may be observed on adrenal CT. If the patient wants to pursue the surgical option, a more accurate subtype test than adrenal CT is required. It is essential for clinicians to understand that normalization of blood pressure is not the only goal. The treatment objectives for patients with PA include resolution of hypokalemia as well as prevention of morbidity and mortality associated with hypertension, progressive chronic kidney disease, and further cardiovascular damage. The cause of PA dictates the optimal treatment option.

Answers

Q1. (d)
Q2. (a), (c)
Q3. (a), (b)
Q4. (a), (b), (c), (d), (e)

References

1. Young WF Jr. Diagnosis and treatment of primary aldosteronism: practical clinical perspectives. J Intern Med. 2019;285:126–48.
2. Kayser SC, Dekkers T, Groenewoud HJ, et al. Study heterogeneity and estimation of prevalence of primary aldosteronism: a systematic review and meta-regression analysis. J Clin Endocrinol Metab. 2016;101:2826–35.
3. Douma S, Petidis K, Doumas M, et al. Prevalence of primary hyperaldosteronism in resistant hypertension: a retrospective observational study. Lancet. 2008;371:1921–6.
4. Calhoun DA, Nishizaka MK, Zaman MA, Thakkar RB, Weissmann P. Hyperaldosteronism among black and white subjects with resistant hypertension. Hypertension. 2002;40:892–6.
5. Ahmed AH, Gordon RD, Ward G, et al. Effect of combined hormonal replacement therapy on the aldosterone/renin ratio in postmenopausal women. J Clin Endocrinol Metab. 2017;102:2329–34.
6. Funder JW, Carey RM, Mantero F, et al. The management of primary aldosteronism: case detection, diagnosis, and treatment: an endocrine society clinical practice guideline. J Clin Endocrinol Metab. 2016;101:1889–916.
7. Lim V, Guo Q, Grant CS, et al. Accuracy of adrenal imaging and adrenal venous sampling in predicting surgical cure of primary aldosteronism. J Clin Endocrinol Metab. 2014;99:2712–9.
8. Umakoshi H, Ogasawara T, Takeda Y, et al. Accuracy of adrenal computed tomography in predicting the unilateral subtype in young patients with hypokalemia and elevation of aldosterone in primary aldosteronism. Clin Endocrinol. 2018;88:645–51.

A 12-Year-Old Boy with Recurrent Carpopedal Spasm

15

Introduction

This case scenario beautifully exemplifies the classic evolution of autoimmune diseases in polyglandular endocrinopathy, namely APS type-1. The entity was initially diagnosed as a schoolboy presenting with recurrent mucocutaneous candidiasis, hypocalcemic tetany, and nail dystrophy. He was well on medication, though later lost to follow-up. Few years later, he was readmitted with hypotensive shock and recurrent tetany which occurred on discontinuation of drugs. Interestingly, his plasma ACTH was very high, and cortisol was low, thereby unmasking an adrenal crisis. During the illness, he developed intestinal tuberculosis, and anti-TB drugs might have precipitated the adrenal crisis.

Case Scenario

A 12-year-old schoolboy presented with intermittent diarrhea and oral thrush since infancy with recurrent episodes of spasm of hands over the last 6 years. The boy was the second issue of a consanguineous marriage having normal developmental milestones. He complained about the frequent passage of watery, voluminous stools associated with cramping lower abdominal pain. He experienced occasional low-grade fever and undocumented weight loss during this period. He used to develop a painful, whitish patch over the oral mucosa on and off again, which would temporarily improve with the use of a topical antifungal agent. He also noticed frequent periodical and acral tingling and numbness, followed by carpopedal spasm and at times generalized convulsion. Such an episode used to be terminated by giving I/v injection in hospital (could not mention the name of injection).

S. Jahan, *Endocrinology Case Studies*,
https://doi.org/10.1007/978-981-95-2770-0_15

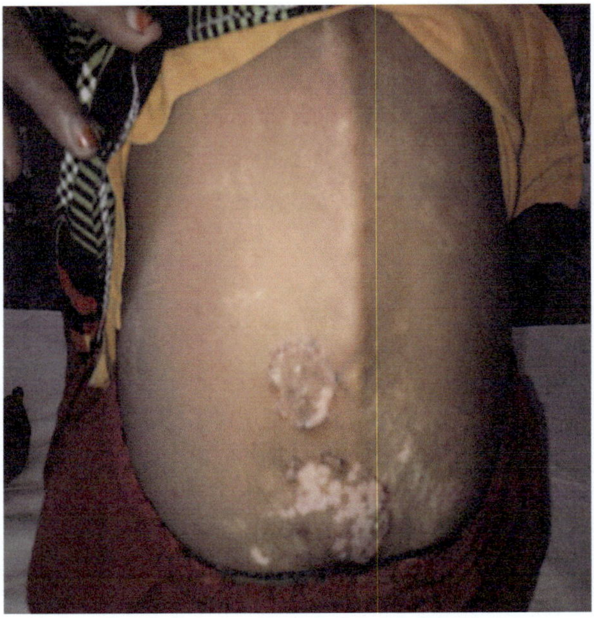

Fig. 15.1 Patches of vitiligo over back

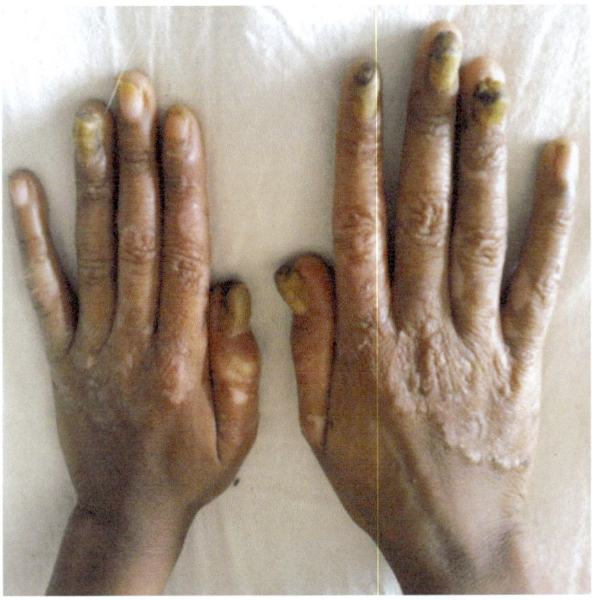

Fig. 15.2 Nail dystrophy and fungal infection

He had stunted growth (height < third percentile) with mild anemia, BMI—11.6 kg/m^2, pulse—70 beats/min, regular, and BP—80/60 mmHg. There were patches of vitiligo and widespread fungal infection all over his body. He had mucocutaneous candidiasis and features of nail dystrophy (Figs. 15.1 and 15.2).

Q1. What is the probable diagnosis depending on clinical features?
(a) Autoimmune poly-endocrinopathy type-2 (APS type-2)
(b) Autoimmune poly-endocrinopathy type-1(APS type-1)
(c) IPEX syndrome
(d) Idiopathic hypoparathyroidism
(e) None of the above

APS type-1, also known as APECED (autoimmune poly-endocrinopathy, candidiasis, and ectodermal dystrophy) tends to occur in the first decade and has three major components, namely mucocutaneous candidiasis, autoimmune hypoparathyroidism, and Addison disease. It is often associated with ectodermal dystrophy, e.g., abnormal nails and teeth. On the other hand, APS type-2 affects middle-aged females more commonly. It is most often characterized by autoimmune thyroid disease, Addison disease, and type-1 DM. IPEX is a rare X-linked disorder of infancy associated with severe malabsorptive diarrhea resulting from autoimmune enteropathy, immune dysfunction manifested by eczema, neonatal DM, and immune thrombocytopenia. If untreated, it is often fatal in early infancy. The boy had a florid picture of APS type-1 since infancy, except for the presence of adrenal insufficiency. So, he must have had APS type-1. Even though he had recurrent diarrhea, the absence of other cardinal features of IPEX made it an unlikely possibility in him.

Q2. What might be the cause of recurrent diarrhea in this boy?
(a) Celiac disease
(b) Cystic fibrosis
(c) Pancreatic insufficiency
(d) Intestinal candidiasis
(e) Lymphangiectasia

APS type-1 results from recessive mutation of the AIRE gene causing widespread immune intolerance with multisystemic involvement. GIT manifestation might include all the abovementioned diseases as well as gut enzyme deficiency like CCK deficiency, or diarrhea itself could be simply idiopathic. In this case, investigations revealed intestinal tuberculosis that is often endemic in this part of the world, and treatment with anti-TB medication led to partial resolution of diarrhea.

Investigations showed mild microcytic hypochromic anemia with high ESR; fecal fat estimation revealed steatorrhea; and the hormonal panel was suggestive of primary hypoparathyroidism with an absence of any other hormonal axis involvement (Table 15.1).

He was put on elemental calcium, active vitamin D supplements, topical antifungal, and miconazole that led to immediate response. For the next 1 year, he was doing well on regular follow-up. Biochemical screening of his parents and siblings failed to reveal any hormonal abnormality related to APS type-1. He soon had a rapid catch-up growth with the appearance of pubertal signs.

Table 15.1 Biochemical profile of the patient at initial admission showing primary hypoparathyroidism

Tests	Result	Reference value
RBS	4.5 mg/dL	–
Serum calcium	8.5 mg/dL	8.5–10
Serum albumin	38 gm/L	34–54 mg/dL
Serum inorganic phosphate	6.8 mg/dL	4–7 mg/dL
Serum PTH	< 3 pg/mL	11–67 pg/mL
Serum TSH	2.15 μIU/mL	0.4–4 μIU/mL
Serum basal cortisol	1630.5 nmol/L	138–690 nmol/L
Serum vit B12	424.5 pgm/mL	239–931 pg/mL
Anti-TG antibody	<20.1 IU/mL	116 IU/mL
Tissue transglutaminase IgA	13.5 U/ mL	≤3 U/mL

Q3. What are the biochemical parameters to be followed up in this case?
   (a) Basal and co-syntropin stimulated plasma cortisol
   (b) Serum FT4 and TSH
   (c) Autoantibodies against interferon-ω and interferon-α
   (d) Antithyroid antibody (anti-TPO and anti-TG Ab)
   (e) NALP5 antibody

While following up, APS type-1 hormonal test including calcium panel and adrenal function should be done on an annual basis. Autoantibodies against interferon-ω and interferon-α have >95% specificity for APS type-1, and among the organ-sensitive antibodies, NALP5 is a marker of parathyroid involvement by autoimmunity. So, tests for these antibodies if available should be done at 3–5 years intervals. On the other hand, autoimmune thyroiditis is not commonly associated with type-1 APS, and thyroperoxidase antibodies are rather nonspecific. So, they are not routinely recommended.

The boy was lost to follow-up 1 year later. Two years later, he was admitted again in a critical state. He had lost considerable weight and had profound anorexia and dysphagia. On examination, he was drowsy, hyperpigmented, and hypotensive having upper abdominal tenderness. During the examination, he had an episode of convulsion necessitating emergency management.

Q4. What might be the reason behind this clinical deterioration?
   (a) Adrenal crisis
   (b) Hypocalcemic tetany
   (c) GIT malignancy
   (d) Pituitary apoplexy
   (e) Diabetic ketoacidosis

His parents admitted that they had discontinued the prescribed drugs for the last few months. He presented with hypocalcemic tetany that was treated with I/V

Table 15.2 Biochemical report at second admission suggestive of primary hypoparathyroidism and hypoadrenalism

Tests	Result	Reference value
RBS	3.9 mg/dL	–
Serum calcium	6.3 mg/dL	8.5–10
Serum albumin	32 g/L	34–54 g/L
Serum inorganic phosphate	7.2 mg/dL	4-7 mg/dL
Serum basal cortisol	**55 nmol/L**	138–690 nmol/L
Plasma ACTH	**1230 pg/mL**	0–46 pg/mL

calcium gluconate. Interestingly, this time, his basal cortisol was found to be only 55 nmo/L and plasma ACTH was 1230 pg/mL. So, he was in an acute adrenal crisis in addition to hypocalcemic tetany. He was immediately started on I/V hydrocortisone and given volume correction with normal saline. Though DKA can sometimes occur in such cases as patients with APS type-1 are prone to develop type-1 DM, and profound weight loss or acute abdomen might be the presenting symptom, however, hypoglycemia instead of hyperglycemia excluded the diagnosis (Table 15.2). Recurrent oropharyngeal candidiasis sometimes leads to esophageal carcinoma causing dysphagia and weight loss in APS type-1. After resuscitation, he underwent UGIT endoscopy revealing extensive esophageal candidiasis that resolved with systemic antifungal therapy. At present, he is on regular follow-up and doing well.

Discussion

APS type-1, also known as autoimmune poly-endocrinopathy–candidiasis ectodermal dystrophy (APECED), is characterized by the classic triad of mucocutaneous candidiasis, autoimmune hypoparathyroidism, and Addison's disease, which form three of the most common components of the disorder. Mutations of the autoimmune regulator gene, AIRE, located on the short arm of chromosome 21 (near markers D21s49 and D21s171 on 21p22.3), are associated with APS-1 [1]. APS type-1 is autosomal recessive in inheritance. APS-1 is characteristically recognized in early childhood. Infants can present with chronic or recurrent mucocutaneous candidiasis in the first year of life, followed by hypoparathyroidism and Addison's disease, but new components can develop at any age. Recurrent candidiasis commonly affects the mouth and nails and, less frequently, the skin and esophagus. Chronic oral candidiasis can result in atrophic disease with areas suggestive of leukoplakia. Ectodermal dystrophy is another component of the syndrome (manifested by pitted nails, keratopathy, and enamel hypoplasia) and cannot be attributed to hypoparathyroidism. The diagnosis of APS-1 is highly likely when two or more of the primary component disorders (i.e., mucocutaneous candidiasis, hypoparathyroidism, and Addison's disease) are present. Autoantibodies against interferon-α and interferon-ω have been identified in almost 100% of subjects with APS-1, regardless of age at screening. These are specific for APS type-1 and recommended for screening if available. Screening to allow the early detection of new disorders before overt symptoms and signs develop is recommended, including autoantibody studies, electrolytes, calcium and phosphorus levels, thyroid and liver function tests, blood sugar, and plasma vitamin B12. Patients at risk for adrenal failure can be screened by the measurement of basal ACTH and supine plasma renin activity (PRA), followed by dynamic testing as appropriate. Treatment includes calcium and active vitamin D supplementation, steroid replacement, and antifungal medicine for hypoparathyroidism, adrenal insufficiency, and mucocutaneous candidiasis, respectively.

Answer

Q1. (a), (b)
Q2. (a), (b), (c), (d), (e)
Q3. (a), (c), (d)
Q4. (a), (b)

Reference

1. Barker JM, et al. The immunoendocrinopathy syndromes. In: Melmed S, et al., editors. William's textbook of endocrinology. 12th ed. Philadelphia: Elsevier Saunders; 2011. p. 1773–5. John F. Kennedy Blvd.

A 60-Year-Old Male Presented with Bilateral Painful Gynecomastia

16

Introduction

This case scenario discusses one of the rarest endocrine paraneoplastic syndromes in renal cell carcinoma, namely gynecomastia. An elderly male presented with rapidly progressive tender gynecomastia with weight loss was evaluated for an underlying etiology. He was found to have very high β-HCG and CEA. USG-guided FNAC of the kidney revealed metastatic renal cell carcinoma staining positive for β-HCG. He developed rapidly deteriorating neurologic deficit resulting from CNS metastatic deposit and succumbed to death shortly thereafter.

Case Scenario

A 60-year-old, retired cultivator presented with bilateral breast swelling for the last 3 months, followed by a change in voice over the past 20 days. The breast swelling was insidious in onset, symmetrical, and slowly increasing in size without any discharge and became painful about 20 days back. He did not give a history of any offending drug, symptoms related to hypogonadism, testicular swelling, and chronic liver or kidney disease. He also complained of a recent change in voice which became indistinct, lacking modulation with imprecise consonants though with intact comprehension. On further query, he gave a history of constipation which had been persistent for the past 5 months with no rectal bleeding or pain. He also complained of anorexia with unintentional weight loss. He did not give any history of cough, chest pain, hematemesis, melena, fever, and night sweats. He was normotensive, nondiabetic, and was an ex-smoker with 20 pack years.

On examination, he was found to be cachectic, anemic, and had tender bilateral gynecomastia with colorless, watery discharge from both nipples (Fig. 16.1). He

S. Jahan, *Endocrinology Case Studies*,
https://doi.org/10.1007/978-981-95-2770-0_16

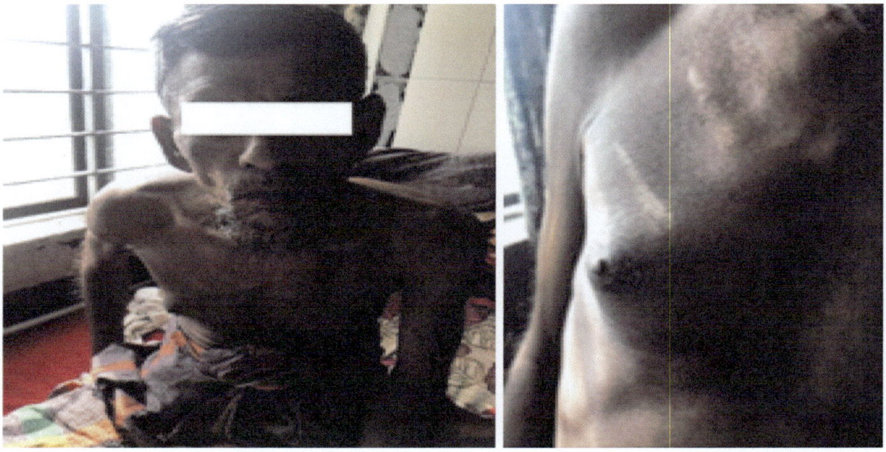

Fig. 16.1 Patient presenting with cachexia and tender gynecomastia

also had slurred speech, anosmia, right-sided hemiparesis, and vesicular breath sound with prolonged expiration on chest auscultation. There was no lymphadenopathy, clubbing, abdominal organomegaly, ascites, or any testicular mass.

Q1. What are the features suggestive of underlying malicious diagnosis in this case?
- (a) Old age
- (b) Bilateral concentric breast bud
- (c) Rapid, painful breast enlargement
- (d) Anorexia with significant weight loss
- (e) Long history of smoking

Gynecomastia is rather a common phenomenon in old age owing to increased adiposity with age (adipose tissue is a major site of aromatization of androgens to estrogens), decreased serum free testosterone due to aging (with decreased testosterone production as well as increased binding of testosterone to SHBG), and greater use of medications that may alter androgen or estrogen concentrations or action. Asymmetric, eccentric breast bud fixed to an underlying structure having retracted nipple with bleeding or nipple discharge, ulceration, or associated lymphadenopathy is very much suggestive of carcinoma of the breast. On the other hand, bilateral concentric breast bud with watery/milky nipple discharge indicates hormonal imbalance as in this case. However, the tender gynecomastia of short duration in the context of marked weight loss and anorexia pointed toward a malicious etiology. Bronchial carcinoma in a smoker can be associated with a plethora of paraneoplastic syndrome including gynecomastia.

Routine investigations revealed microcytic hypochromic anemia with markedly elevated ESR. SGPT and serum creatinine were normal, excluding any chronic kidney or liver disease. Then, hormonal investigations were undertaken accordingly (Table 16.1).

Table 16.1 Biochemical investigation of the patient showing markedly elevated β-HCG and CEA

Parameter	Patient's value	Reference range
ALT (U/L)	32	14–63 IU/L
Creatinine (mg/dL)	0.77	0.5–1.3 mg/dL
FGB (mmol/L)	5.1	≤5.5 mmo/L
Albumin (gm/L)	38	34–54 mg/dL
β HCG (mIU/L)	**102239.70**	0.00–2.5 miu/L
Testosterone (ng/dL)	434.63	270–1734 ng/dL
Estrogen (pg/mL)	**358.24**	10–40 pg/mL
AFP (ng/mL)	1.58	10–20 ng/mL
CEA (U/mL)	**13.9**	0–2.5 ng/mL
CA 19–9 (ng/mL)	4.9	0–37 U/mL

Q2. What hormone testing is necessary in this patient?
(a) LH, FSH, and testosterone
(b) DHT, androstenedione, and DHEA
(c) Estradiol and prolactin
(d) B-HCG
(e) Alpha-fetoprotein

Low testosterone levels with low/low-normal LH and FSH suggest hypogonadotropic hypogonadism, whereas high LH and FSH levels indicate hypergonadotropic hypogonadism. Elevated circulating thyroid hormone sometimes may lead to gynecomastia due to increased aromatase activity. Hyperprolactinemia may be responsible for galactorrhea with gynecomastia by inducing central hypogonadism and creating an imbalance between estrogen and testosterone; androstenedione and DHEA are various types of gonadal and adrenal androgens, and they are not included in the hormonal work-up of gynecomastia, rather they are essential in the evaluation of other disorders like hyperandrogenism, disorder of sexual differentiation, adrenal disease, etc. Alpha-fetoprotein is an oncofetal antigen expressed in hepatocellular carcinoma that is not directly implicated in the causation of gynecomastia. Human chorionic gonadotropin (β-hCG)-secreting tumor of gonadal or extragonadal germ-cell origin (also called eutopic hCG production) or a tumor derived from a nontrophoblastic tissue, such as a large-cell carcinoma of the lung or some gastric or renal-cell carcinomas (also called ectopic hCG production), may secrete estradiol indirectly through the stimulatory effects of LH receptors on Leydig cells.

Our patient had markedly elevated β-HCG level with estradiol that was inappropriately high for a male. So, he was likely to have a β-HCG-secreting tumor. For the localization of such a tumor, we searched for eutopic and ectopic sources. CXR P/A view revealed widespread cannonball opacities suggestive of secondaries (Fig. 16.2). USG of scrotum showed bilateral normal-appearing testes with few enlarged inguinal lymph nodes. USG of the abdomen revealed multiple SOLs in the liver and a (1.5 × 1.6) cm lesion in the upper pole of the right kidney (Figs. 16.3 and 16.4). USG-guided FNAC of hepatic and renal

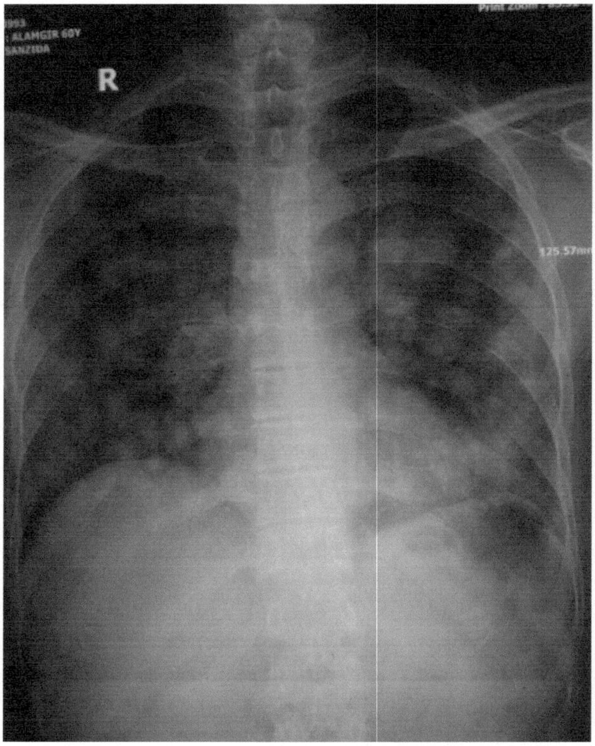

Fig. 16.2 Chest X-ray showing cannonball opacities in both lungs

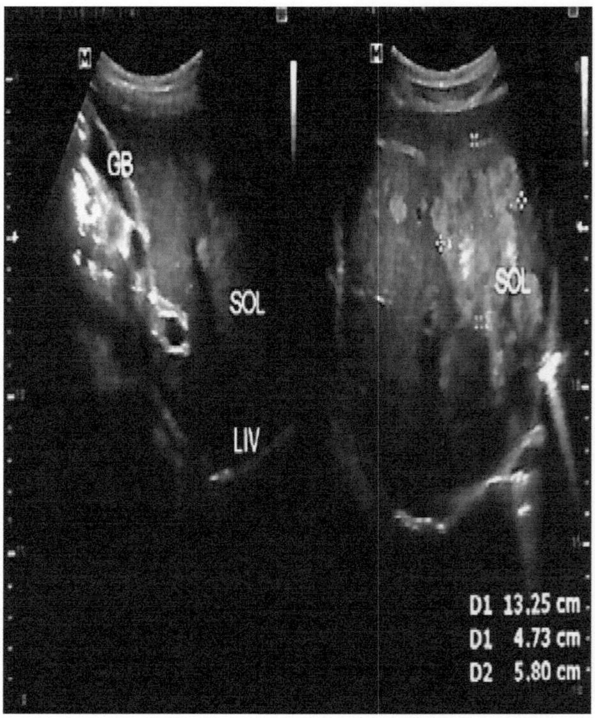

Fig. 16.3 USG showing echogenic mass in right kidney

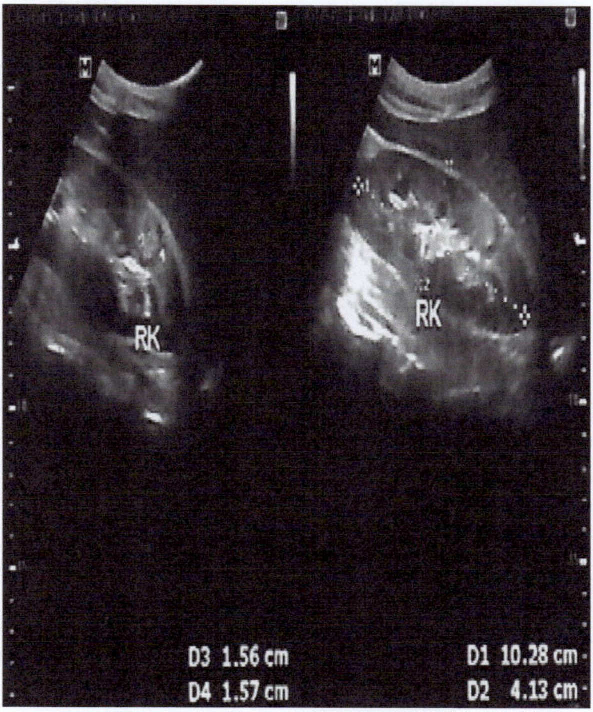

Fig. 16.4 Multiple SOLs in liver on abdominal USG

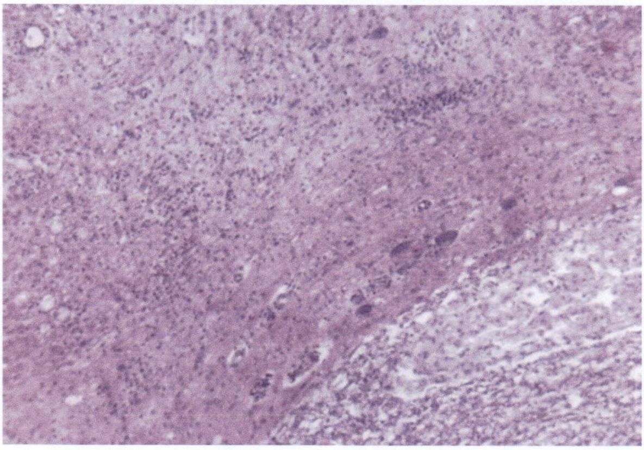

Fig. 16.5 FNAC from right kidney showing malignant renal cells. (*Source: https://phil.cdc.gov/Details.aspx?pid=2762&utm)

SOLs was advised, and it showed metastatic adenocarcinoma of liver and renal cell carcinoma which stained positive for β-HCG (Fig. 16.5). So, the final diagnosis was metastatic renal cell carcinoma with paraneoplastic gynecomastia.

Q3. Which endocrinopathy does not commonly occur as a paraneoplastic syndrome?
- (a) Cushing syndrome
- (b) Conn syndrome
- (c) Humoral hypercalcemia of malignancy
- (d) SIADH
- (e) Pyrexia

Paraneoplastic endocrine syndromes (PNS) mostly result from the production of bioactive substances from the neoplastic cells of endocrine or neuroendocrine origin that are widely distributed throughout the lungs, gastrointestinal tract, pancreas, thyroid gland, adrenal medulla, skin, prostate, and breasts [1, 2]. Bronchial carcinoma is most associated with endocrine PNS, e.g., squamous cell carcinoma with hypercalcemia and small-cell carcinoma (SCLC), and carcinoid with Cushing syndrome and SIADH. Ectopic aldosterone secretion occurring from a tumor has not been described in the literature. Pyrexia can occur due to the liberation of cytokines from pheochromocytoma, paraganglioma, and fibrous angioid histiocytoma. On the other hand, RCC is an uncommon source of paraneoplastic syndrome, though it has been associated with hypercalcemia, hypertension, polycythemia, Cushing syndrome, and alterations in glucose metabolism with anemia, neuropathy, pyrexia, and amyloidosis among the nonendocrine presentations [3, 4].

During the hospital stay, the patient had rapid neurologic deterioration in terms of right hemiparesis and slurring of speech. He was rapidly shifted to the oncology unit where he was started on palliative chemotherapy. However, he succumbed to death soon afterward.

Q4. What might be the reason behind his neurologic manifestation?
- (a) Paraneoplastic syndrome
- (b) Cerebral metastasis
- (c) Stroke
- (d) Functional symptoms
- (e) None of the above

Urgent CT scan of the brain done before transferring to the oncology department showed diffuse metastasis to the left frontoparietal area that was producing the neurologic deficit. He had a very high β-HCG level which was associated with widespread metastasis and poor prognosis. It is postulated that it may act via repression of the transforming growth factor-β (TGF-β) receptor complex, inhibiting tumor cell apoptosis. β-hCG in serum is a poor prognostic factor independent of the stage and grade of the tumor that can be used to identify a subgroup of patients with an increased risk of aggressive disease.

Discussion

The paraneoplastic syndromes (PNS) comprise a fascinating array of symptom complexes that are manifested systemically in association with a wide variety of malignant (or less commonly benign) tumors [1]. The symptoms are unrelated to the local effects of the tumors or their distant spread but are the consequence of the production of hormones, growth factors, cytokines, or other substances by the tumor cells [2–6]. Clinical presentations may be dominated by the presence of otherwise unexplained symptoms related to hormone overproduction or may be manifested by symptoms relating to the neuromuscular, integumentary, osteoarticular, hematopoietic, vascular, renal, or other systems. Interestingly, paraneoplastic syndromes can antedate the clinical appearance of malignancy by months to years. Most paraneoplastic endocrine syndromes are related to the production of peptide hormones. ***Renal cell carcinoma (RCC)*** is unique among the genitourinary malignancies in that close to one-third of affected patients show signs and symptoms of paraneoplastic syndrome. The paraneoplastic syndromes associated with renal cell carcinoma range from those manifesting in constitutional symptoms (i.e., fever, cachexia, and weight loss) to those that result in specific metabolic and biochemical abnormalities (i.e., hypercalcemia, nonmetastatic hepatic dysfunction, amyloidosis, gynecomastia, etc.) [7]. β-HCG, normally made by the syncytiotrophoblastic cells of the placenta, has been found in elevated levels in patients with RCC [8]. Elevated levels of this hormone in any adult male or nonpregnant female should suggest malignancy. High levels of β-HCG have been documented in up to 6% of those patients with RCC [9]. It is hypothesized that HCG may act via repression of the transforming growth factor-β (TGF-β) receptor complex, inhibiting tumor cell apoptosis and explaining its association with metastasis and poor prognosis [10]. Elevated levels of free β-subunit of HCG have also been reported as an adverse prognostic factor in renal cell cancer, independent of clinical and histological grade [11]. As in many other paraneoplastic syndromes, β-HCG levels decline following nephrectomy but may rise with the disease recurrence [10].

Answer

Q1. (c), (d), (e)
Q2. (a), (c), (d)
Q3. (b)
Q4. (b)

References

1. Agarwala SS. Paraneoplastic syndromes. Med Clin North Am. 1996;80:173–84.
2. Baylin SB, Mendelsohn G. Ectopic (inappropriate) hormone production by tumors: mechanisms involved and the biological and clinical implications. Endocr Rev. 1980;1:45–77.
3. Gagel RF. Endocrine manifestations of tumors: "ectopic" hormone production. In: Goldman L, editor. Cecil textbook of medicine. On-line edition. Philadelphia, PA: WB Saunders; 2000.
4. Johnson B. Paraneoplastic endocrine syndromes (Part 6, Section 1, Chapter 100). In: Braunwald E, Fauci AS, Isseuracker KJ, et al., editors. Harrison's principles of internal medicine (on-line). New York, NY: McGraw-Hill; 2001–2003.
5. Strewler GJ. Humoral manifestations of malignancy. In: Larsen PR, Kronenberg HM, Melmed S, Polonsky KS, editors. Williams textbook of endocrinology. 10th ed. Philadelphia, PA: WB Saunders; 2003. p. 1834–2573.
6. Turner HE, Wass JAH. Ectopic hormone syndromes. In: DeGroot LJ, Jameson JL, editors. Endocrinology. 4th ed. Philadelphia, PA: WB Saunders; 2001. p. 2559–73.
7. Palapattu GS, Kristo B, Rajfer J. Paraneoplastic syndromes in urologic malignancy: the many faces of renal cell carcinoma. Rev Urol. 2002;4(4):163–70.
8. Braunstein GD, Vaitukaitis JL, Carbone PP, et al. Ectopic production of human chorionic gonadotropin by neoplasms. Intern Med. 1973;78:39–45.
9. Kuida C, Braunstein GD, Shintaku P, et al. Human chorionic gonadotropin expression in lung, breast and renal carcinomas. Arch Pathol Lab Med. 1988;112:282–5.
10. Yatura S, Harrara E, Nopajaroonsri C, Singal R, Gill S. Gynaecomastia attributable to human chorionic gonadotrophin-secreting giant cell carcinoma of lung. Endocr Pract. 2003;9:2335.
11. Hotakainen K, Ljungberg B, Paju A, Rasmuson T, Alfthan H, Stenman U-H. The free b-subunit of human chorionic gonadotropin as a prognostic factor in renal cell carcinoma. Br J Cancer. 2002;86:1859.

A 51-Year-Old Male with Severe Uncontrolled HTN and Unilateral Adrenal Mass

17

Introduction

This case scenario presents the dilemma in the diagnosis of severe, resistant HTN in a 51-year-old schoolteacher. He had several episodes of hypertensive emergencies presenting with hypokalemia on one hand and one episode of hyperkalemia following the initiation of ARB on the other hand. His radiologic imaging studies including adrenal CT scan and DTPA renogram revealed renal artery stenosis and an adrenal incidentaloma. After a thorough clinical and biochemical evaluation coupled with imaging analysis, he was diagnosed of having ectopic pheochromocytoma with renal artery stenosis, a rare combination leading to severe and resistant HTN.

Case Scenario

A 51-year-old schoolteacher was diagnosed with HTN and IHD 12 years back. At that time, he presented with occasional exertional chest discomfort, headache, palpitation, and sweating. His HTN remained relatively well-controlled over the following year. However, he developed severe HTN during the next 11 years that was resistant to beta-blockers, CCB, ARB, and diuretics. He already had developed target organ damage as evidenced by proteinuria, progressively rising serum creatinine, and LVH. On several occasions, he needed emergency hospitalization presented with severe HTN, headache, and vomiting with hyponatremia and hypokalemia. Once he developed AKI on CKD precipitated by pyelonephritis and was treated conservatively. For evaluation of renal involvement, he underwent USG of W/A at that time showing a smaller kidney on the left side with gross parenchymal insufficiency confirmed on DTPA renogram. He was further evaluated with a CT of abdomen showing a small ovoid hypodense soft tissue lesion (0.88 × 0.71 cm) in the

S. Jahan, *Endocrinology Case Studies*,
https://doi.org/10.1007/978-981-95-2770-0_17

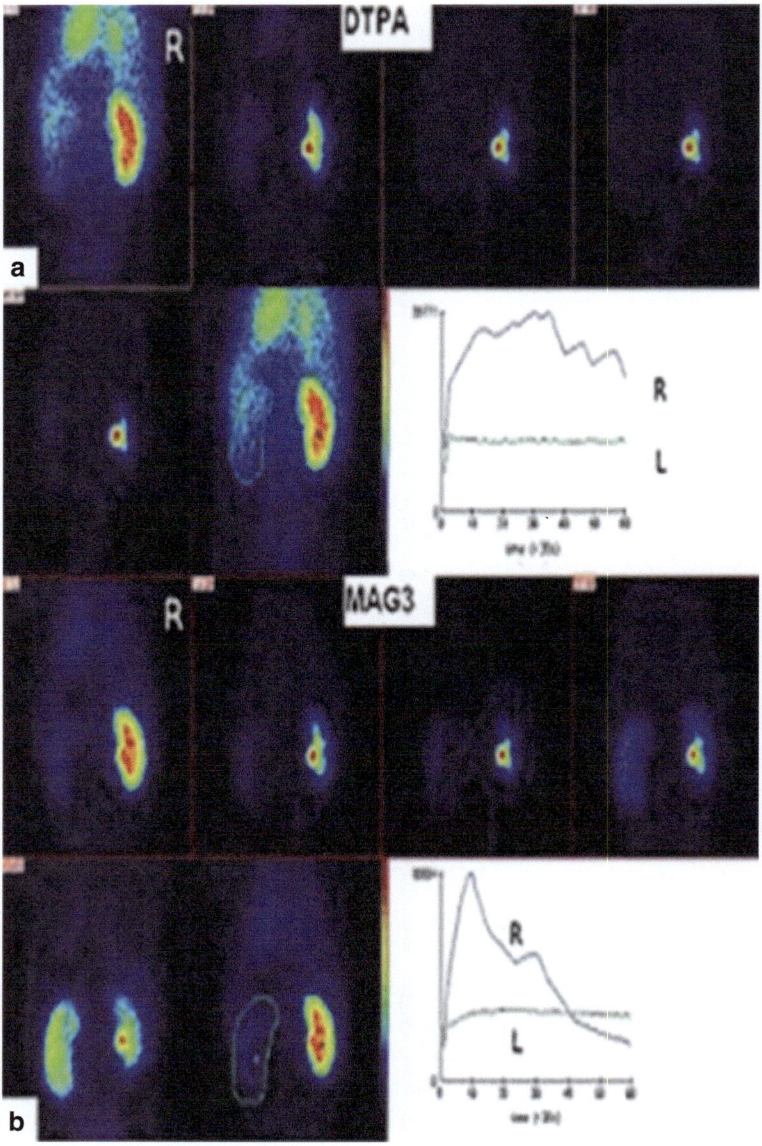

Fig. 17.1 DTPA renogram (panel **a** and **b**) showing gross parenchymal insufficiency of left kidney

left adrenal region, exhibiting a mild homogeneous enhancement on contrast scan with a mild hypoplastic/atrophic left kidney (Figs. 17.1 and 17.2). Finally, he was admitted into the endocrine unit for further evaluation.

On examination, he was mildly anemic having a BMI of 19.3 kg/m^2, pulse—78/min, regular, and moderate-to-severe HTN was noticed on several occasions with no postural drop. Other physical examination findings including fundoscopy were unremarkable.

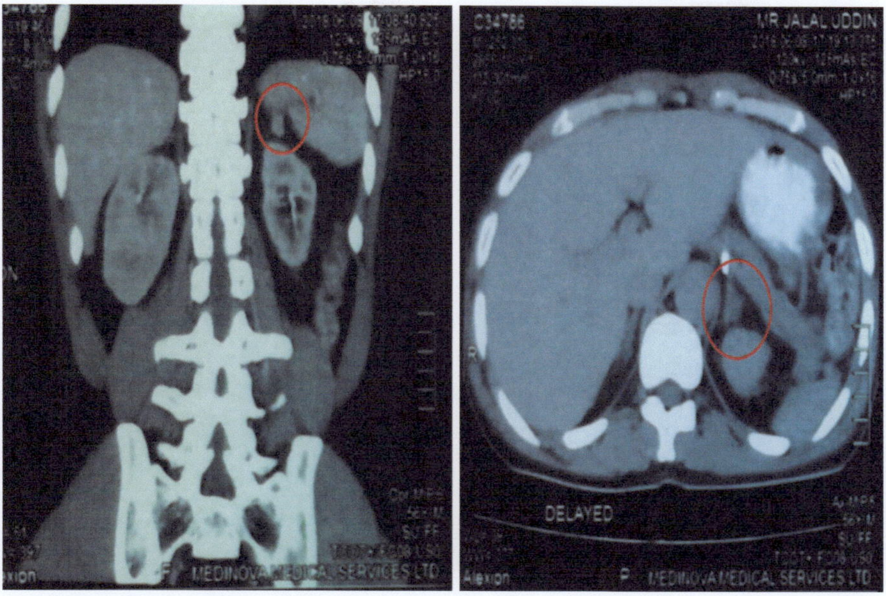

Fig. 17.2 Small left adrenal mass

Q1. What is the possible diagnosis based on the above history?
 (a) Pheochromocytoma
 (b) Conn syndrome
 (c) Renal artery stenosis (RAS)
 (d) Essential HTN with adrenal incidentaloma
 (e) Adrenal Cushing syndrome

Adrenal Cushing was unlikely because the stigmata of Cushing syndrome was completely absent. He had few episodes of hypokalemia along with hyponatremia, making Conn a possibility. However, hypokalemia on each occasion occurred after an episode of severe vomiting, and he was on diuretic most of the time. Interestingly, once after the initiation of ARB, he developed hyperkalemia, and the left kidney was found to be smaller in size on USG, making RAS a likely possibility, though it would not account for the adrenal lesion. Young-onset, severe HTN with an adrenal mass would be much too of a coincidence, making essential HTN unlikely. On the other hand, pheochromocytoma could be a possibility in the case of young-onset, severe, resistant HTN with an adrenal tumor where the episodes of hospitalization with headache and vomiting could be attributed to Pheo spell.

He was investigated accordingly. Routine investigations including serum electrolytes were normal except for mildly raised serum creatinine. Hormonal investigations to exclude functioning adrenal tumors were also negative. Only serum metanephrine was at the upper limit of normal (Table 17.1).

Table 17.1 Biochemical report of the patient showing apparently normal adrenal hormonal level

Tests	Result	Reference range
24 h urinary free cortisol	63.56	50–190 µg/24 h
24 h urinary metanephrine	1502.50	<1775 nmol/day
DHEA-S	224.90	38–313 µg/dL
Plasma aldosterone	7.1	3–40 ng/dL
Renin concentration	1.5	0.4–3 ng/dL

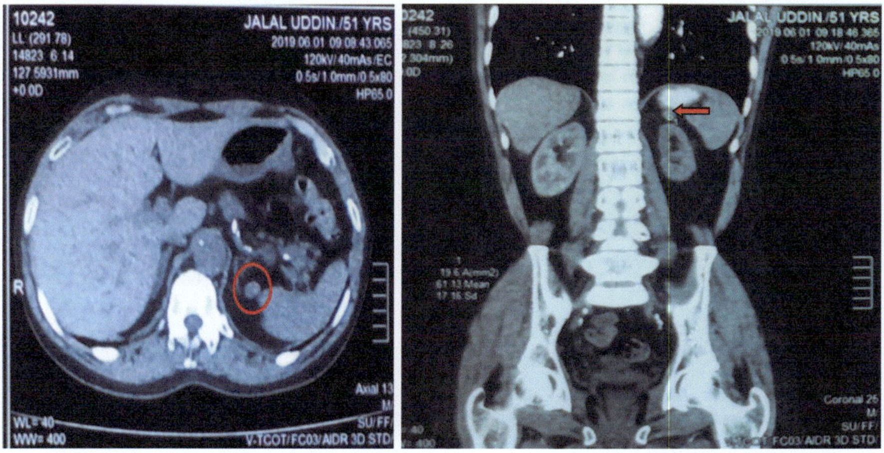

Fig. 17.3 Repeat CT adrenal (6 years after the initial one) showing an enlarged left adrenal lesion (red circle and arrow)

Doppler USG of kidneys showed small, echogenic left kidney (6.1 cm) having a tardus parvus pattern in intrarenal branches suggestive of >70% stenosis in the left renal artery. MRA of kidneys also found severe stenosis of the left renal artery with nonfunctioning kidney on that side. Then, CT of abdomen was repeated in the adrenal protocol that showed a significant growth of previous adrenal tumor now measuring (1.1 × 1.3) cm, ovoid, and hypodense in nature having a HU of+45 and AWR (absolute washout ratio) of 59%. Interestingly, it was located lateral to the limb of the left adrenal gland (Fig. 17.3).

Q2. What is the diagnosis based on imaging phenotype on repeat scan?
(a) Conn tumor
(b) Adrenocortical carcinoma (ACC)
(c) Adrenal metastasis
(d) Pheochromocytoma
(e) Nonfunctioning adrenal adenoma

The lipid attenuation value of the X-ray beam (expressed in HU) in a CT scan is most useful for the characterization of a tumor. Benign tumors like Conn adenoma generally have a HU <10 U as they are rich in cholesterol. However, in 30% of such cases, the lipid-poor adenoma may be present having higher attenuation values. On the other hand, ACC, metastasis, and pheochromocytoma tend to have HU >20 U. The AWR of benign tumor is >50%, whereas, in malignancy, pheochromocytoma is usually <50% at 15 min. So, considering these facts, benign functioning/nonfunctioning adenoma like Conn was excluded. ACC is commonly >4 cm, irregular with a fuzzy margin, and heterogeneous with the areas of necrosis and hemorrhage. When functionally active, it often co-secretes multiple hormones like cortisol and DHEA and tends to follow a rapid downhill course. Adrenal metastasis occurs in the context of known primary malignancy, often bilateral with irregular margin and functionally inactive. So, these were unlikely diagnoses, though the imaging characteristics were partially suggestive. The ultimate diagnosis was pheochromocytoma considering the high pretest probability on the clinical background and typical imaging phenotype with a growth rate of >0.8 cm/year. However, its slightly unusual location suggested an ectopic pheochromocytoma with renal artery stenosis.

Q3. Which of the following statements is true?
- (a) RAS is a mere coincidence in pheochromocytoma.
- (b) Pheochromocytoma can lead to RAS.
- (c) RAS can lead to pheochromocytoma.
- (d) Treatment of pheochromocytoma leads to resolution of RAS.
- (e) Treatment of RAS is enough for normalization of BP.

It is reported that in 7% of cases, pheochromocytoma has been associated with RAS. The reasons are multiple like compression of an intrarenal branch of a renal artery by an ectopic pheochromocytoma, organic fibrosis of ipsilateral renal artery by a persistently elevated level of catecholamine in the territory, and as a component of neuroectodermal syndrome, e.g., NF1. It can be a separate entity as well [1–4]. When the stenosis is mild, the removal of the tumor often reverses the condition. If the stenosis is >70%, then the stenosis needs to be corrected separately for the restoration of normal BP. In our patient, the stenosis was severe with grossly nonfunctional left kidney unilateral nephrectomy, with adrenalectomy done with adequate perioperative preparation.

Q4. What important complications might occur during surgery of this patient?
- (a) Hypertensive crisis
- (b) Hyperkalemia
- (c) Hypotensive shock
- (d) Adrenal crisis
- (e) Acute renal shutdown

Perioperative preparation of pheochromocytoma is very crucial lest there might be a catastrophe from hypertensive crisis or hypotensive shock. Prior treatment with sequential α- and β-blockers with liberal salt intake is essential to prevent a hypertensive emergency.However, it might occur and need to be treated with I/V labetalol, GTN, or nitroprusside. On the other hand, removal of the tumor without prior volume resuscitation might lead to hypotensive shock that should be corrected with pressors, colloids, etc. Hyperkalemia is the complication that arises after the removal of Conn's adenoma, especially in patients who had severe hypokalemia prior to surgery. Adrenal crisis does not occur after pheochromocytoma removal as it does not suppress the HPA axis. In the case of bilateral adrenalectomy offered for bilateral Pheo, there is every possibility of adrenal crisis.

The intraoperative period was uneventful. The histopathology report of the resected tumor was suggestive of benign pheochromocytoma having a typical Zell Ballen appearance. Afterward, the patient BP was well controlled with a single antihypertensive, and he remained free of symptoms on subsequent follow-up.

Discussion

Pheochromocytoma should be suspected in patients who have one or more of these features like hyperadrenergic spells, early-onset severe or resistant hypertension, a positive family history of pheochromocytoma or familial syndrome that predisposes to catecholamine-secreting tumors (e.g., MEN 2, NF1, and VHL), an incidentally discovered adrenal mass, concurrent hypertension and diabetes, or H/O pressor response during anesthesia and surgery [5]. In a hypertensive patient with adrenal incidentaloma as in the previous case, it is mandatory to rule out pheochromocytoma. The imaging phenotype is often suggestive if not diagnostic of pheochromocytoma. Imaging characteristics consistent with a benign cortical adenoma include round and homogeneous density, smooth contour with sharp margination, diameter usually less than 3 cm, unilateral location, and low unenhanced CT attenuation values (<10 HU) with rapid contrast washout at 10 min [6, 7]. The imaging phenotype consistent with pheochromocytoma includes enhancement with an intravenous contrast medium on CT, high signal intensity on T2-weighted MRI, cystic and hemorrhagic changes, and variable size; also, the tumor may be bilateral in 10% cases. The treatment of choice for pheochromocytoma is complete surgical resection. Surgical survival rates are 98–100% and are highly dependent on the skill of the endocrinologist, endocrine surgeon, and anesthesiologist team [8]. In rare instances, pheochromocytoma may be associated with ipsilateral ***renal artery stenosis (RAS)*** which accentuates the HTN. If the RAS is severe, then it might require separate surgical correction for the optimum outcome of the patient.

Answer

Q1. (a), (c)
Q2. (d)
Q3. (b)
Q4. (a), (c)

References

1. Gill IS, Meraney AM, Bravo EL, Novick AC. Pheochromocytoma coexisting with renal artery lesions. J Urol. 2000;164:296–301.
2. Kuzmanovska D, Sahpazova E, Kocova M, Damjavnoski M, Popov Z. Haeochromocytoma associated with reversible renal artery stenosis. Nephrol Dial Transplant. 2001;16:2092–4.
3. Chotsampancharoen T, Patrapinyokul S, Reegkling C, Vachvanichsanong P. Impaired differential renal function in a child with pheochromocytoma. J Hum Hypertens. 2005;19:751–4.
4. Camberos A, Bautista N, Rubenzik M, Applebaum H. Renal artery stenosis and pheochromocytoma: coexistence and treatment. J Pediatr Surg. 2000;35:714–6.
5. Endocrine HTN, William F Young, chapter 16, William's textbook of endocrinology, 12th edition.
6. Korobkin M, Brodeur FJ, Francis IR, et al. CT time: attenuation washout curves of adrenal adenomas and nonadenomas. AJR Am J Roentgenol. 1998;170:747–52.
7. Szolar DH, Korobkin M, Reittner P, et al. Adrenocortical carcinomas and adrenal pheochromocytomas: mass and enhancement loss evaluation at delayed contrast-enhanced CT. Radiology. 2005;234:479–85.
8. Plouin PF, Duclos JM, Soppelsa F, et al. Factors associated with perioperative morbidity and mortality in patients with pheochromocytoma: analysis of 165 operations at a single center. J Clin Endocrinol Metab. 2001;86:1480–6.

A 27-Year-Old Female with Primary Amenorrhea and an Inguinal Lump

18

Introduction

This is one of the few cases of sexual ambiguity included in this book depicting an apparent healthy young female, evaluated for primary amenorrhea, who was found to have bilateral inguinal lumps and absent uterus/ovaries on pelvic scan. Thorough biochemical investigations, followed by detailed anatomic imaging, suggested a case of 46 XY DSD resulting from Leydig cell hypoplasia (LCH)/testosterone biosynthetic defect. The final diagnosis was established by a histopathology report of the resected inguinal testes typical of LCH, though her genetic mutational analysis was inconclusive. Thereafter, she underwent bilateral orchidectomy and subsequent estrogenization.

Case Scenario

A 27-year-old healthy-looking unmarried female presented with the complaint of failure to establish spontaneous menstruation till date. She was the second issue of a consanguineous marriage. Her mother stated that she was born in term uneventfully by NVD and her antenatal period was unremarkable. They noticed seemingly normal female external genitalia at birth, and she was given a female gender assignment. Her subsequent neonatal and childhood period was uneventful with age-appropriate developmental milestones. However, pubertal changes were somewhat delayed, with adrenarche evident at the age of 14 years. She noticed an acceleration of her height in comparison to her peers at the age of 17 years. Her breast development was poor till date, and menarche never occurred. She gave no history suggestive of any adrenal insufficiency. Her family history was insignificant with the absence of any genital ambiguity, unexpected changes at puberty, or early death of

S. Jahan, *Endocrinology Case Studies*,
https://doi.org/10.1007/978-981-95-2770-0_18

infants within her relatives. Her mother recalled that her elder sister had a herniorrhaphy at 3 years of age with the removal of an ovoid structure from the hernial sac. At present, this elder sister is married though infertile. Our patient had a USG of the pelvis done a few years back showing the absence of uterus or ovaries.

On examination, she appeared well cooperative though a little depressed, and her psychological sex orientation was that of a female. She was taller in comparison to her peers having a eunuchoid body habitus. Sexual maturation rating by Tanner staging revealed apparently elevated breast without a disc, poorly developed areola with retracted nipple (B2), and sparse hair growth in the axilla and pubis (P2). She had normal-looking labia majora, fused labia minora, presence of normal urethral but blind vaginal orifice, and neither clitoromegaly nor scrotum/penis were visible. Interestingly two palpable lumps (2 × 2) cm, firm, smooth, and nontender were found, one in each inguinal region.

Q1. What are the probabilities based on clinical features?
- (a) Mullerian agenesis
- (b) Mosaic Turner syndrome
- (c) Complete androgen insensitivity syndrome (CAIS)
- (d) Leydig cell hypoplasia (LCH)
- (e) Swyer syndrome

Considering the scenario of female phenotype with absent Mullerian structures, our initial diagnosis was Mullerian agenesis, popularly known as Myers–Rokitansky–Kuster–Hauser syndrome (MRKHS), but the presence of palpable lumps in the inguinal canal indicative of testes made us think otherwise. Mosaic Turner syndrome might present with sexual infantilism with absent gonads on imaging even in the absence of any physical stigmata. On the other hand, Swyer syndrome/46XY pure gonadal dysgenesis could have been a probability considering the relatively underdeveloped female phenotype with agonadism on pelvic USG. However, the palpable testes excluded these possibilities altogether. Now, she represented a case of 46XY DSD or under-virilized male, and the differentials included complete androgen insensitivity syndrome (CAIS), defect in testosterone biosynthetic pathway, or Leydig cell hypoplasia. A rare possibility would have been ovotesticular DSD (46XX/XY), where the ovotestes sometimes descend below the inguinal ligament.

We planned our investigations accordingly. The routine biochemical parameters including serum electrolytes were within normal limits. The USG of W/A showed an enlarged right kidney (14.8 × 8.1 cm) having multiple cysts of variable sizes with the loss of architecture. Uterus and both ovaries were not visualized in their location. Both the testes were seen in the inguino-vulvar region, right testis measuring about (24 × 10) mm and left testis (24 × 10) mm. After color Doppler application, flow was seen in both the testes (Fig. 18.1). Her karyotype, as we expected, proved to be 46XY. Hormonal evaluation was done, which showed the relatively normal level of FSH and AMH (Table 18.1), whereas serum LH was very high with a very low level of testosterone.

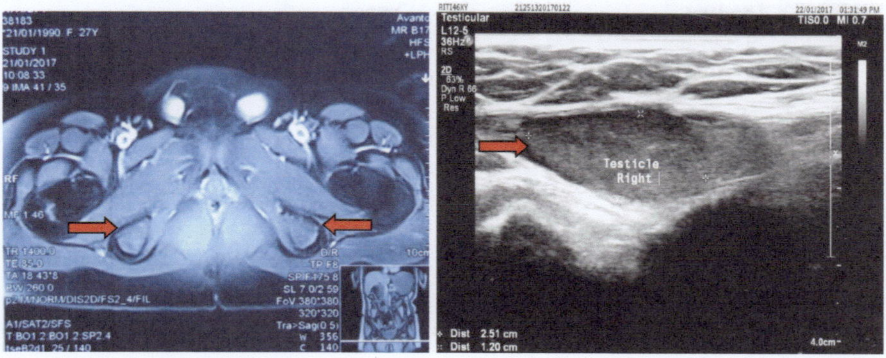

Fig. 18.1 MRI and USG of pelvis/ingunal region showing bilateral ovoid structure indicative of testes

Table 18.1 Hormonal report of the patient showing low plasma testosterone and estrogen with inappropriately high LH

Parameter	Patient's value	Reference value
FSH	6.1 mIU/mL	1.6–8.0 mIU/mL
LH	**37.9 mIU/mL**	1.5–9.3 mIU/mL
Serum testosterone	<1 ng/mL	25–110 ng/mL
Serum estradiol	13 pg/mL	0–56 pg/mL
Plasma ACTH	15 pg/mL	0–46 pg/mL
Basal cortisol	356 nmol/L	138–690 nmol/L
DHEA-S	227.80 μg/dL	110–510 μg/dL
Serum AMH	233 pmol/L	>200 pml/L

Q2. How the viability of inguinal testes can be determined?
- (a) Color Doppler USG of testes
- (b) Anti-Mullerian hormone (AMH)
- (c) HCG stimulation test
- (d) GnRH stimulation test
- (e) Inhibin B

Though most of the time baseline normal level of testosterone with normal/low-normal LH is enough to predict the viability of testes, at other times, dynamic tests or surrogate biomarkers might be required. AMH is one such marker with a cutoff >200 pmol/L indicating functioning Sertoli cell, as in our case. Inhibin B is secreted by granulosa cells of the ovarian follicle and is a surrogate marker of ovarian reserve, useful for the evaluation of female hypogonadism or infertility. The presence of flow in testes on Doppler USG color is a gross assessment of its viability. However, in the absence of hormonal activity, it is considered insignificant. GnRH stimulation tests with the GnRH/GnRH analog are frequently done for the evaluation of delayed or precocious puberty. The HCG stimulation test is specially designed for the assessment of testicular viability. HCG at a dose of 1500 IU I/M is given for

Table 18.2 HCG stimulation test showing insignificant increment of serum testosterone/DHT following HCG injection

Before stimulation	Result	Reference range
Serum testosterone	**0.46 ng/mL**	270–1734 ng/mL
Serum DHT	**0.34 ng/mL**	14–77 ng/mL
17(OH progesterone)	1.9	0.5–2.1 ng/mL
After stimulation		
Serum testosterone	**0.17 ng/mL**	NA
DHT	**0.65 ng/mL**	NA
17(OH progesterone)	2.16 ng/mL	0.5–2.1 ng/mL

subsequent 3 days. One day after the last injection, blood is assayed for LH, FSH, serum testosterone, and DHT, and 24 h urine is collected for measuring C19/20 keto steroids or testosterone precursors.

Q3. What is the interpretation of the HCG stimulation test in this case?
- (a) Androgen insensitivity syndrome (CAIS)
- (b) 5 -α-reductase deficiency syndrome
- (c) Leydig cell hypoplasia (LCH)
- (d) Primary testicular failure (POF)
- (e) Testosterone biosynthetic block

HCG stimulation test showed a negligible increment of serum testosterone level, which under normal circumstances is expected to be 2–20 folds higher (Table 18.2). Ancillary tests of adrenal functions were found to be normal that commonly are abnormal in the enzymatic block of steroid synthesis (Table 18.1). In CAIS, we expect high LH, FSH, and testosterone, both before and after HCG stimulation. Also, such patients tend to be well feminized resulting from high estrogen derived from the aromatization of testosterone. In 5-α-reductase deficiency, there is pubertal virilization in a phenotypic female, and T: DHT following HCG stimulation is typically >10:1. Primary gonadal failure occurring in early intauterine life may lead to a complete female phenotype. However, the gonads would be absent in such a case with the preservation of Mullerian structures, which was quite the opposite in her. Our patient had relatively normal AMH and FSH levels indicative of spared Sertoli cell function, thereby biochemically excluding gonadal agenesis too. Ultimately, we narrowed down our differentials into LH receptor resistance/Leydig cell hypoplasia or enzymatic defect in the testosterone biosynthetic pathway. Normal adrenal function and failure of buildup of steroid precursors on the HCG stimulation test excluded the testosterone biosynthetic defect.

At this stage, we consulted urologists for definitive management, and according to their advice, MRI of the abdomen was done that showed two T1 intermediate and T2 hyperintense areas measuring about (1.86 × 1.79) and (2.36 × 2.18) cm in the right and left inguinal canals, respectively—possibly undescended testis as well as rudimentary prostate at the bladder base along with gross right-sided

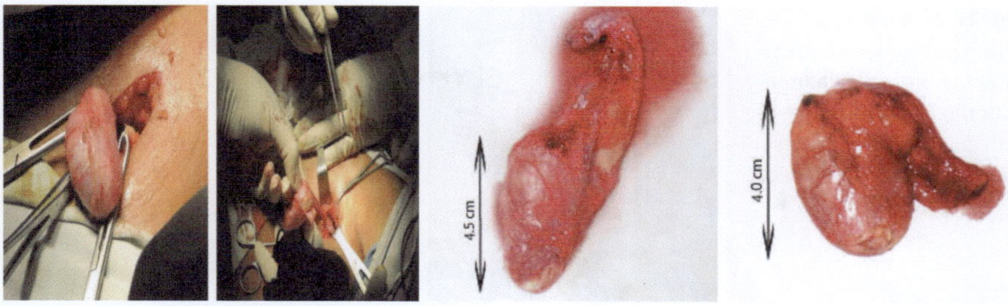

Fig. 18.2 The removed testis on orchidectomy with the attached epididymis and spermatic cord

hydronephrosis with PUJ obstruction. Following this comprehensive workup, she underwent bilateral orchidectomy with vaginoplasty and right-sided pyeloplasty/stenting. On gross anatomy, the right testis was (4 × 2.5) cm with attached epididymis (2.8 cm) and spermatic cord (3.8 cm). The cut surface of the testis and epididymis was unremarkable, while the left testis was (4.5 × 2.5) cm with the attached epididymis and spermatic cord (3.0 cm and 5.5 cm, respectively) (Fig. 18.2). The ultimate diagnosis was uncovered as the histopathologic report became available. The report showed atrophic changes with an increased number of Sertoli cells, absence of spermatogenic cells or Leydig cells, and no granuloma or malignant cells were evident; so, the diagnosis was compatible with Leydig cell aplasia/hypoplasia (Figs. 18.3 and 18.4).

Q4. Which of the following statement regarding LCH is true?
   (a) Leydig cell hypoplasia (LCH) is an X-linked disorder.
   (b) Female sex conversion is mandatory in LCH.
   (c) Genetic mutation analysis is necessary for confirmatory diagnosis.
   (d) Transmission rate into her offspring is 25%.
   (e) None of the above.

LCH is a rare autosomal recessive disorder characterized by an inability of the body to respond to LH, which is normally responsible for signaling Leydig cells of the testicles to produce testosterone and other androgen sex hormones. It is caused by genetic mutations in LH-CGR, a gene that encodes the LH/hCG receptor. The DNA sequencing of the patient's LH-CGR gene did not reveal any mutations in its coding sequences. However, the analysis did not exclude the possibility of mutations in the introns, promoter regions, or other supporting genes rather than LH-CGR (Fig. 18.5). So, genetic mutation analysis cannot establish the diagnosis in all cases as no abnormality might be detected in a handful of cases. LCH is of two types, the first one presenting with male phenotype with incomplete virilization and the second type presenting as complete female phenotype. In the latter, female sex conversion is usually the rule with corrective surgery and estrogen therapy. The transmission rate for the autosomal recessive disorder is 50%. She will not be able to procreate as her gonads were removed, which showed no evidence of spermatogenesis.

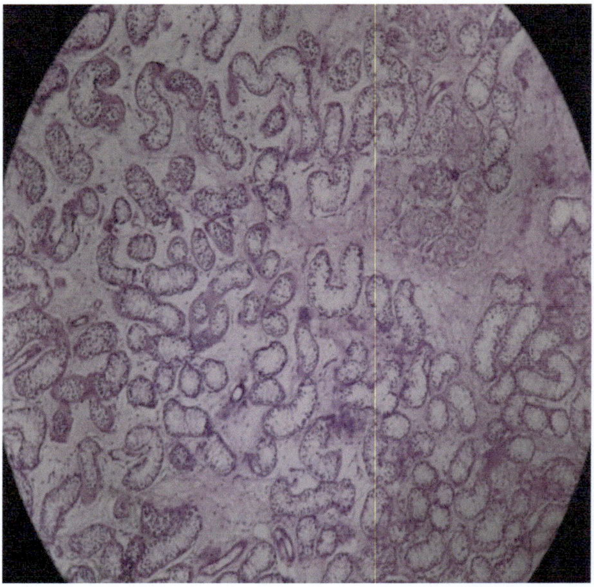

Fig. 18.3 The histopathology specimen showing atrophic change with the absence of Leydig cells and spermatogenesis and increased Sertoli cells

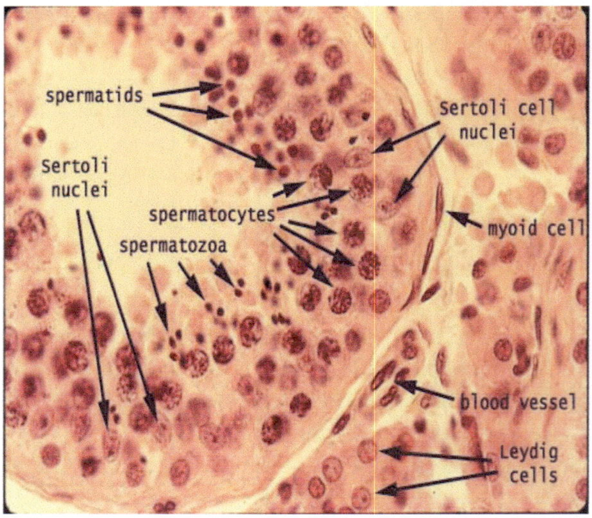

Fig. 18.4 Histopathology of normal testis

After orchiectomy and vaginal dilatation, she was estrogenized with medication. Her diagnosis and its impact on her fertility were only disclosed to the patient herself. Recently, she got married and is living a happy conjugal life. All of her family members are undergoing genetic screening.

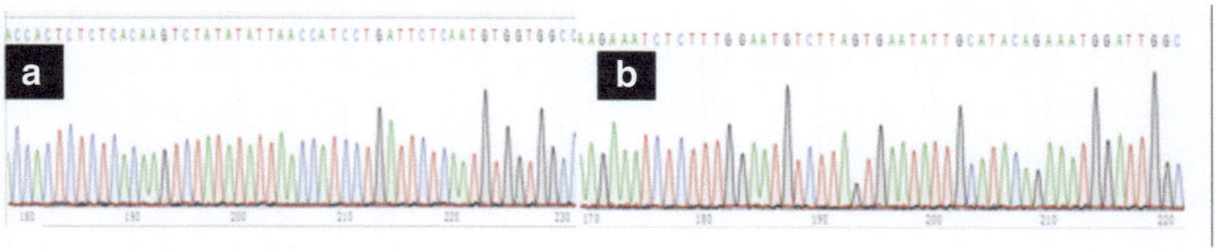

Fig. 18.5 DNA sequencing (panel **a** and **b**) of the patient's genes for luteinizing hormone and chorionic gonadotropin failed to reveal any mutations in the coding regions

Discussion

The term "***disorders of sex development***" ***(DSD)*** is proposed to define congenital conditions in which a disharmony between chromosomal, gonadal, and anatomical sex exists [1]. Human sex development is a highly complex process controlled by multiple genes and hormones. Sexual differentiation follows a sequence of events with sequential development of the gonads, internal genital ducts, and external genitalia [2]. Abnormalities of sexual differentiation may occur at any step along the way producing predictable clinical syndromes. In 2006, the LWPES and the ESPE published a consensus statement on the management of "intersex" disorders and proposed a new classification for DSD [3, 4]. This includes three main diagnostic categories: sex chromosome DSD, 46XY DSD (formerly male pseudohermaphroditism / under-virilized male), and 46 XX DSD (formerly female pseudohermaphroditism). The 46 XY DSD are characterized by ambiguous or female external genitalia, caused by incomplete intrauterine masculinization and the presence or absence of Mullerian structures. ***Leydig cell aplasia (or hypoplasia) (LCH)***, also known as Leydig cell agenesis, is a rare autosomal recessive genetic and endocrine syndrome affecting an estimated 1 in 1,000,000 biological males. It is a unique example of 46XY DSD characterized by an inability of the body to respond to LH, which is normally responsible for signaling Leydig cells of the testicles to produce testosterone and other androgen sex hormones. It is caused by genetic mutations in LH-CGR, a gene that encodes the LH/hCG receptor [5]. LH normally acts through the LH/hCG receptor to stimulate the growth of Leydig cells in the testicles and the production of androgens such as testosterone and DHT. In Leydig cell hypoplasia/aplasia, however, there is a reduced capacity for the LH/hCG receptor to respond to LH. This results in hypoplasia or absence of Leydig cells, testicular atrophy, and lower-than-normal androgen levels. These patients are characterized by ambiguous or female external genitalia and the presence or absence of Mullerian structures. Complete absence of virilization results in the normal female external genitalia, and they generally seek medical attention at pubertal age, due to the absence of breast development and/or primary amenorrhea. Over 20 inactivating mutations of LHCGR have been identified to date, scattered throughout the gene [6, 7]. Even then, there are a handful of case reports and case series where none was identified. However, most of the patients with the clinical phenotype of LCH did not have causative mutations, suggesting that changes in other regions of the LH receptor gene, such as the large introns or the promoter region, may be responsible for most cases. Alternatively, the displayed phenotype may be the result of other genetic defects. This further underscores the importance of thorough clinical analysis of patients before molecular analysis of a particular gene is performed [8].

Answer Key

Q1. (d)
Q2. (a), (b), (c)
Q3. (c), (e)
Q4. (e)

References

1. Berenbaum SA, Bailey JM. Effects on gender identity of prenatal androgens and genital appearance. Evidence from girls with congenital adrenal hyperplasia. J Clin Endocrinol Metab. 2003;88(3):1102–6.
2. Sax L. How common is intersex? A response to Anne Fausto- Sterling. J Sex Res. 2002;39:174–8.
3. Hughes IA, Houk CP, Ahmed SF, Lee PA. LWPES Consensus Group; ESPE Consensus Group. Consensus statement on management of intersex disorders. Arch Dis Child. 2006;91:554–63.
4. Summary of Consensus Statement on Intersex Disorders and Their Management. Christopher P. Houk, MD, PhDa, Ieuan A. Hughes, FMedSci, FRCPCH b, c, S. Faisal Ahmed, FRCPCHd, Peter A. Lee, MD, PhDe, PEDIATRICS Volume 118, Number 2, August 2006.
5. Berthezene F, Forest MG, Grimaud JA, et al. Leydig-cell agenesis: a cause of male pseudohermaphroditism. N Engl J Med. 1976;295:969–72.
6. Arnhold IJ, de Mendonça BB, Toledo SP, et al. Leydig cell hypoplasia causing male pseudohermaphroditism: case report and review of the literature. Rev Hosp Clin Fac Med Sao Paulo. 1987;42:227–32.
7. Martens JW, Lumbroso S, Verhoef-Post M, et al. Mutant luteinizing hormone receptors in a compound heterozygous patient with complete Leydig cell hypoplasia: abnormal processing causes signaling deficiency. J Clin Endocrinol Metab. 2002;87:2506–13.
8. Richter-Unruh A, Martens JW, Verhoef-Post M, et al. Leydig cell hypoplasia: cases with new mutations, new polymorphisms and cases without mutations in the luteinizing hormone receptor gene. Clin Endocrinol. 2002;56:103–12.

A 44-Year-Old Male with Proximal Myopathy, Polyuria, and Polydipsia

19

Introduction

This is the story of a middle-aged man who was evaluated for uncontrolled HTN and debilitating proximal myopathy of recent onset. He had a H/O surgery for invasive olfactory neuroblastoma (ONB) several years earlier; he took no further treatment post-surgery. His biochemical investigations confirmed the presence of ACTH-dependent Cushing syndrome, though a variable report of ACTH created some confusion initially. Finally, the source was localized following endo-nasal surgery for recurrent olfactory neuroblastoma that stained positive for ACTH. It is one of the few reported cases of ectopic Cushing syndrome due to the paraneoplastic expression of ACTH by ONB.

Case Scenario

A 44-year-old glass factory worker developed debilitating proximal myopathy, progressive in nature and without any diurnal variation over the last 7 months. He also noticed edema appearing first in the dependent parts and later spreading on to other areas of the body. There was no H/O orthopnea, breathlessness, or paroxysmal nocturnal dyspnea. He complained about polyuria, polydipsia, and significant weight loss of about 15 kg in the last 4 months. He had been hypertensive for the last 3 years on amlodipine and Olmesartan 5/20 mg daily, with poor compliance and poor control of blood pressure. He had functional endoscopic sinus surgery (FESS) 3 years back for a left sinus mass that presented with epistaxis. Biopsy of the mass revealed a poorly differentiated carcinoma, and IHC confirmed olfactory neuroblastoma (ON). External beam radiotherapy was advised as further treatment, but he declined to take it. In recent times, his proximal weakness hampered his functional abilities to the extent of necessitating immediate hospitalization into our unit.

S. Jahan, *Endocrinology Case Studies*,
https://doi.org/10.1007/978-981-95-2770-0_19

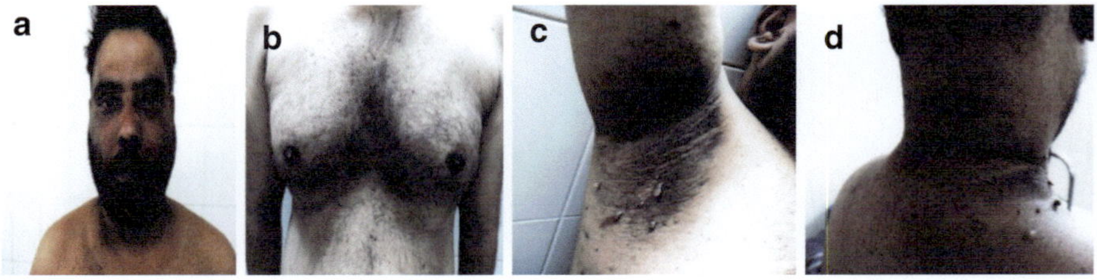

Fig. 19.1 (**a**) Patient with plethoric moon face, (**b**) bilateral gynecomastia, and (**c** and **d**) acanthosis and skin tag in the axilla and neck

Fig. 19.2 Skin thinning and bruises over fingers

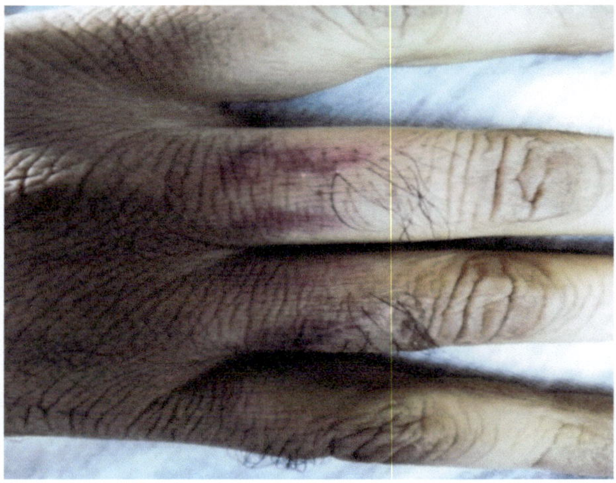

Physical examination revealed a man with florid Cushing syndrome. He had a plethoric moon face, central obesity with dorsocervical fat pad, and fullness of supraclavicular fossa. Acanthosis nigricans was present in the axilla and nape of the neck, along with prominent skin tags. There was significant proximal myopathy, both in the upper and lower limbs. He also had skin thinning, spontaneous bruises, onychomycosis, and bilateral gynecomastia (Figs. 19.1, 19.2, and 19.3). However, the wide purple striae characteristic of Cushing syndrome was absent.

After admission, he was investigated thoroughly. Routine biochemical investigations showed neutrophilic leukocytosis, uncontrolled hyperglycemia, hypertriglyceridemia, and fatty liver; moderate cardiomegaly was evident on CXR. There was hypokalemia on several testing, the lowest one being 1.6 mmol/L; ABG was suggestive of metabolic alkalosis. The thyroid function test and serum calcium panel were normal (Table 19.1).

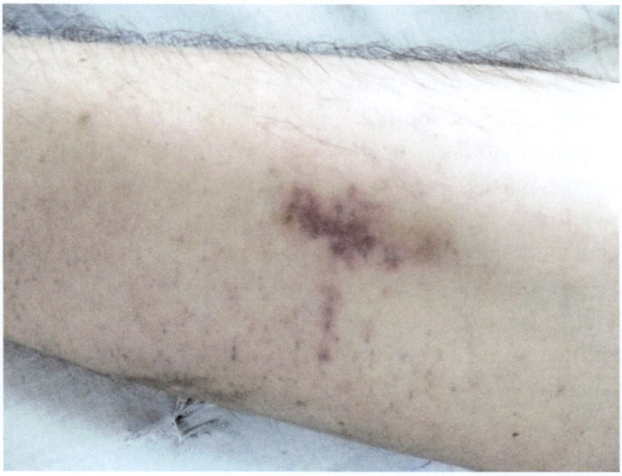

Fig. 19.3 Spontaneous bruises over forearm

Table 19.1 Routine biochemical investigation report of the patient showing hyperglycemia and hypokalemia

Investigations	Results	Reference range
CBC	Hb—13gm/dL, WBC—16,500/mm^3 D/C-N—90%, L—9.0%, M—1.0%, E—0.0%, B—0.0%	Hb—15 ± 2gm/dL, TC of WBC—7 ± 3X10^9/L Platelet—150–450X10^3/l
Plasma glucose (random)	32.2 mmol/L	–
S. Electrolyte	Sodium—141 mmol/L, potassium—2.5 mmol/L, chloride—100 mmol/L, TCO2—29 mmol/L	Na—135–145, K—3.5–5.2, cl—95–107, TCO_2—22–30 mmol/L
SGPT	80 U/L	Up to 40
CPK	95 U/L	24–195
Lipid profile	Total cholesterol—192 mg/dL, HDL-35 mg/dL LDL—109 mg/dl, TG- 239 mg/dl	Up to 200 mg/dL,>45 mg/dl <140 mg/d,50-150 mg/dl
S.Calcium	8.4 mg/dL	8.4–10.4 mg/dl
S.Albumin	34.7 g/L	34–50 gm/L
S. Magnesium	0.89 mmol/L	0.7–1.05 mmol/L
PTH	81.90 pg/mL	7-30 pg/mL

Q1. Ectopic Cushing syndrome is likely in this patient because of the following:
   (a) Proximal myopathy
   (b) Uncontrolled HTN
   (c) Acanthosis nigricans, skin tag
   (d) Edema
   (e) Hypokalemic alkalosis

Ectopic Cushing syndrome comprises about 10% of Cushing syndrome accounting for up to 25% of ACTH-dependent cases. It is more common in males. Half of the cases occur in small-cell carcinoma of lungs and the remainder in a variety of neuroendocrine tumors, rarely in epithelial malignancy. Rapid progression of severe hypercortisolemia is characterized by prominent myopathy and severe HTN with edema caused by the mineralocorticoid action of high circulating cortisol overwhelming 11-β-HSD2 enzyme activity. Weight loss and hyperpigmentation are also common because of underlying malignancy and high ACTH level, respectively. Weight loss in this patient was probably due to uncontrolled DM rather than anything else, and hyperpigmentation was notably absent. Acanthosis nigricans and skin tag indicate underlying insulin resistance that are not features of ectopic Cushing syndrome per se. Hypokalemic alkalosis is present in 90% of such cases, whereas it is found in only 5% of patients with Cushing disease. Other features like striae, bruises, and buffalo hump are poor discriminators between the pituitary and ectopic Cushing syndrome. Cushing disease typically occurs in female between third and fourth decade and commonly presents in an indolent manner predominantly with hyperandrogenic features like hirsutism, oligomenorrhea, or infertility, often with nonspecific headache.

The investigations were carried out to confirm Cushing syndrome and to determine the underlying etiology. It showed high 24 h UFC and salivary cortisol with nonsuppressibility of serum cortisol on low-dose dexamethasone suppression test (LDDST). The report of plasma ACTH created some confusion, being low on two occasions initially, then high on subsequent testing (Table. 19.2).

Table 19.2 Hormonal report of the patient showing nonsuppressed plasma cortisol with variable ACTH levels

Investigations	Results	Reference range
Basal cortisol	**839.6 nmol/L**	101–690 nmol/l
24 h urinary free cortisol	**1249.88 mmol/day**	11.80–485.60 mmol/day
24 h aldosterone in urine	94.40 nmol/day	5.54–60.94 nmol/day
ONDST	**Serumcortisol—649.39 nmol/L**	<50 nmol/L
HDDST	**Serum cortisol—589 nmol/L**	Positive response->50% decrement from baseline
Free cortisol in saliva	**24.10 nmol/L**	1.66–6.90 nmol/L
Serum aldosterone Plasma renin	30.22 ng/dL 0.002 ng/mL	3–40 ng/dL Plasma renin—0.4–3 ng/dL
24 h urinary metanephrine	1551 nmol/L	<1775 nmol/L
Saline loading test	Aldosterone—4 ng/dL	Nonsuppressed >10 ng/dL
DHEAS	245 mcg/dL	38–313 mcg/dL
Plasma ACTH	**5 pg/mL (first occasion)** **10 pg/mL (second occasion)** **124 pg/mL (third occasion)**	8.3–57.8 pg/mL

Q2. What is the reason behind variable ACTH report?
   (a) Storage problem with the sample
   (b) Hook effect
   (c) Heterophile antibody
   (d) ACTH precursor and fragment
   (e) All the above

Following the proper clinical evaluation for pituitary and adrenal disorders, the ultimate diagnosis relies on the accurate measurement of plasma ACTH. It is advised to collect whole blood into lithium heparin or EDTA plastic tubes, taking note of the sample requirement of the immunoassay to be used. The use of inappropriate sample collection containers usually results in abnormally low ACTH concentrations. Glass tubes must be siliconized as ACTH sticks to the glass. Samples should be kept cool and centrifuged within 1 h of collection and the plasma transferred into plastic or siliconized glass tubes. Grossly hemolyzed samples cannot be used. Samples should be transported frozen on dry ice to the reference laboratory. Consequently, special precautions for the handling of blood specimens are widely advocated, and samples are spun immediately, and the plasma is collected and stored at −70 °C until assay. The modern two-site sandwich ACTH immunoassay is a great improvement compared with older methods but still has the potential for interferences such as heterophile antibodies and pro-opiomelanocortin (POMC) and ACTH fragments [1]. More recently, biotin was found to interfere with biotinylated assays, and as the use of over-the-counter substances increases, additional potential interfering substances might be noted [2]. In very rare instances of the ectopic adrenocorticotropic hormone (ACTH) syndrome, the elevated ACTH may be biologically active but not detected by the immunometric assay. However, if a result is inappropriately low, it is important to assess the storage history. Sometimes, very high ACTH in ectopic Cushing syndrome might overwhelm the detection antibody, producing the so-called "Hook effect". ACTH precursor (POMC and CLIP) measurements might provide valuable diagnostic discriminatory evidence of the ectopic ACTH syndrome from Cushing disease [3].

Though the report of ACTH led to confusion, on initial testing, it was low which finally was found to be 124 pg /mL when retested appropriately. So, the patient had ACTH-dependent Cushing syndrome, as evidenced by nonsuppressed serum cortisol with high ACTH. RAAS axis was evaluated for the presence of primary/secondary hyperaldosteronism as he had hypokalemic HTN. However, the presence of florid Cushing syndrome made it unlikely except in the case of the occurrence of adrenal carcinoma which often co-secretes multiple hormones. This was excluded by a negative confirmatory test for hyperaldosteronism, absence of hyperandrogenism, and any adrenal lesion on CT scan.

Q3. Which investigation effectively differentiates between ectopic and pituitary Cushing?
   (a) Pituitary MRI with contrast
   (b) Inferior petrosal sinus sampling (IPSS)

(c) CT of chest and abdomen
(d) Octreoscan
(e) High-dose dexamethasone suppression test (HDDST)

All of the above investigations are done to localize the tumor in ACTH-dependent Cushing syndrome albeit with variable success. Even in the most sophisticated center, MRI is negative in 50% of cases that are later proved to have pituitary microadenoma on surgery. On the other hand, pituitary incidentaloma is relatively common, occurring in about 10% of the general population. Dynamic pituitary MRI can be a reasonable option though not completely reliable. In doubtful cases (normal MRI/microadenoma <6 mm), it is better to perform IPSS that will show lateralization in pituitary cases. When the results of both pituitary MRI and IPSS are negative, then elaborate imaging along with biomarkers for neuroendocrine tumor (NET) needs to be undertaken. On the other hand, even after exhaustive investigations like CT of chest/abdomen or octreoscan, the location of 12–13% of ectopic Cushing will remain elusive. On HDDST, 5% of CD does not suppress, and 5% of ectopic Cushing suppresses. Only IPSS done in centers having the necessary expertise can differentiate between the types of ACTH-dependent Cushing in the most difficult cases.

The serum cortisol remained nonsuppressed on HDDST, possibly indicating ectopic CS. The dynamic pituitary MRI showed only mild pituitary enlargement without any evidence of tumor (Fig. 19.4). There was the presence of a large sino-nasal mass evident on sagittal brain MRI (Fig. 19.5). The adrenals were unremarkable on the CT of abdomen excluding any adrenal etiology (Fig. 19.6). In order to localize the source of high ACTH secretion, CT scans of the chest and abdomen and USG of the neck were done. All were found to be unremarkable. Serum calcitonin was also normal, excluding the possibility of thyroid medullary carcinoma. Ultimately, the diagnosis was clinched after the removal of the recurrent olfactory neuroblastoma. The resected tumor proved positive for ACTH on immunostaining. It also stained positive for other immunomarkers of NET. There was a significant clinical and biochemical improvement on follow-up a few months after the surgery.

Q4. What are the common plasma biomarkers of NET?
(a) Chromogranin A (CgA)
(b) Neuron-specific enolase
(c) Dopamine
(d) 5-hydroxy-indole-acetic acid (5-HIAA)
(e) Ghrelin

NET is characterized by an array of peptide hormones and amines in blood and their metabolites in urine. Chromogranin A /B, neuron-specific enolase, neurotensin Y, and Ghrelin are detectable in plasma, whereas 5-HIAA, homo-vanillic acid (HVA), and dopamine are metabolites for carcinoid tumors and pheochromocytoma in urine.

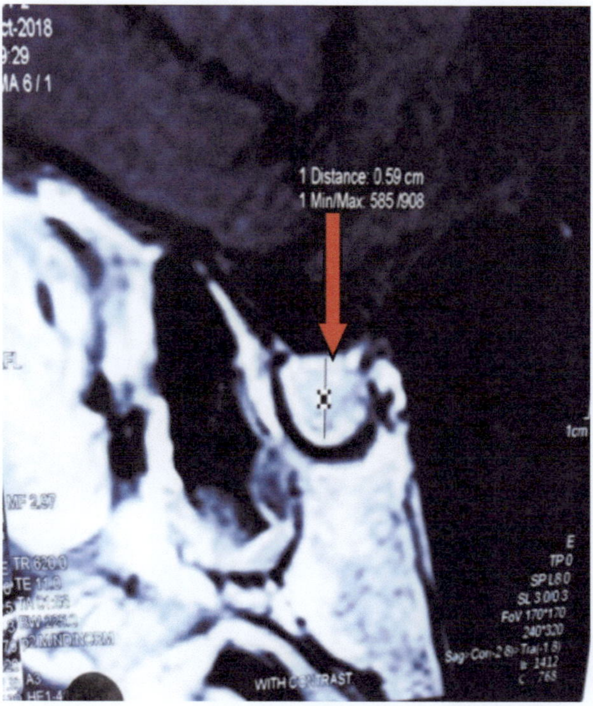

Fig. 19.4 Mild enlargement of the pituitary on dynamic MRI

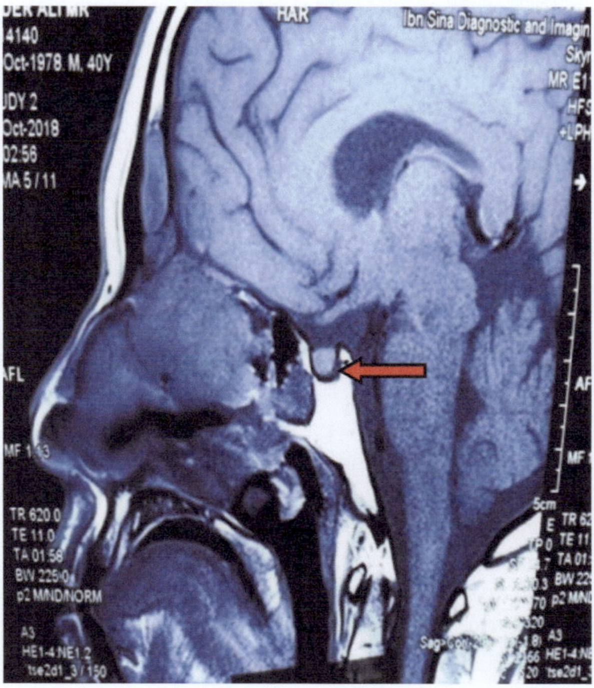

Fig. 19.5 Large sinonasal mass

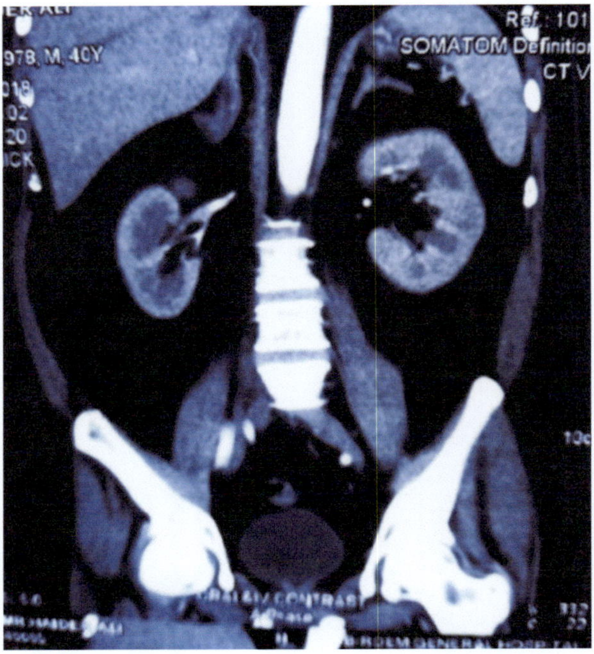

Fig. 19.6 Unremarkable adrenals on CT scan

Discussion

In 15% of cases, Cushing syndrome is associated with nonpituitary tumors secreting ACTH—the ***ectopic ACTH syndrome***. On clinical grounds, these can be divided into two entities: cases occurring in the setting of highly malignant tumors such as small-cell carcinoma of bronchus and more indolent cases occurring in patients with underlying neuroendocrine tumors such as bronchial carcinoids. Circulating ACTH concentrations and cortisol secretion rates can be extremely high. As a result, the duration of symptoms from onset to presentation is short (<3 months). Patients are commonly pigmented, and the metabolic manifestations of glucocorticoid excess are often rapid and progressive. Weight loss, myopathy, and glucose intolerance are prominent symptoms and signs. The association of these features with hypokalemic alkalosis and peripheral edema should alert the clinician toward the diagnosis. After confirmation of Cushing's syndrome clinically and biochemically, the clinician's next step is to determine the cause which relies on the level of plasma ACTH. In Cushing's disease, 50% of patients have a 9 am ACTH level within the normal reference range [9–52 pg/mL]; in the remainder, it is modestly elevated. ACTH levels in the ectopic ACTH syndrome are high [>90 pg/ mL]; nevertheless, overlap values are seen in Cushing's disease in 30% of cases [1]. ***Olfactory [4] neuroblastoma (ONB)*** accounts for 2–3% of tumors of the nasal cavity, and it is likely derived from the olfactory basal reserve cells in the olfactory mucosa [1]. ONB occurs in a wide range of ages and frequently presents with nasal obstruction,

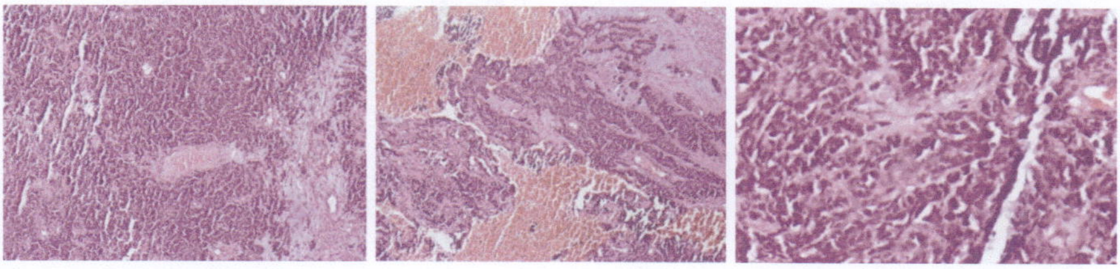

Fig. 19.7 Small-cell neuroendocrine tumor of cervix in hematoxylin and eosin stain

epistaxis, and anosmia. The prognosis of ONB is variable with 10-year survival rates between 45% and 76% [4–6]. Rarely, ONB can cause a paraneoplastic syndrome that is secondary to ectopic secretion of peptides and hormones or an immune response due to cross-reactivity between neoplastic cells and normal tissues [3]. Endocrinological manifestations of ONB include syndrome of inappropriate ADH secretion, ectopic ACTH syndrome, humoral hypercalcemia of malignancy, hypertension caused by catecholamine secretion, and hyperprolactinemia [2, 5, 6]. Only four cases with ACTH secretion confirmed at immunohistochemistry on tissue have previously been reported [1]. Figure 19.7 shows the histopathologic feature of a small-cell neuroendocrine tumor that might present in a similar way with ectopic ACTH syndrome.

Answer

Q1. (a), (b), (d), (e)
Q2. (e)
Q3. (b)
Q4. (a), (b), (e)

References

1. Greene LW, Geer EB, Page-Wilson G, Findling JW, Raff H. Assay-specific spurious ACTH results lead to misdiagnosis, unnecessary testing, and surgical misadventure—a case series. J Endocr Soc. 2019;3(4):763–72.
2. Samarasinghe S, Meah F, Singh V, et al. Biotin interference with routine clinical immunoassays: understand the causes and mitigate the risks. Endocr Pract. 2017;23(8):989–98.
3. Talbot A, Kane JW, White A. Analytical and clinical aspects of adrenocorticotrophin determination. J Ann Clin Biochem. 2003;40:453–71.
4. Onb-Koo BK, An JH, Jeon KH, Choi SH, Cho YM, Jang HC, Chung JH, Lee CH, Lim S. Two cases of ectopic adrenocorticotropic hormone syndrome with olfactory neuroblastoma and literature review. Endocr J. 2008;55:469–75.
5. Kunc M, Gabrych A, Czapiewski P, Sworczak K. Paraneoplastic syndromes in olfactory neuroblastoma. Contemp Oncol (Pozn). 2015;19:6–16.
6. Lacroix A, Feelders RA, Stratakis CA, Nieman LK. Cushing's syndrome. Lancet. 2015;386:913–27.

A 13-Year-Old Boy with Generalized Pigmentation and Fever

20

Introduction

This case scenario elaborately discusses probable differential diagnoses relating to hyperpigmentation, weight loss, and weakness. The schoolboy in the case study initially was thought to have Addison disease on the basis of typical clinical picture and subnormal plasma cortisol level. However, on repeat testing combined with the ACTH stimulation test, he was proved to have an intact HPA axis, effectively ruling out Addison disease. His neurologic symptoms and sensory involvement prompted the investigations to exclude subacute combined degeneration of the spinal cord (SCD) that revealed significant deficiency of serum vit-B12. Finally, the presence of anti-intrinsic factor antibody in blood and gastritis on UGIT endoscopy confirmed the diagnosis of pernicious anemia.

Case Scenario

A 13-year-old schoolboy presented with progressive generalized hyperpigmentation, especially of lips, hands, and feet over the last $1^{1/2}$ years. He also suffered from low-grade fever during the last 6 months, which was intermittent and not documented properly; on query, he gave a history of anorexia with significant weight loss of about 5 kg in the past 3 months along with constipation. He also complained of the passage of three to four episodes of black, tarry stool during the last 2 months. There was no history of nausea, vomiting, dizziness or loss of consciousness, tetany, thyrotoxicosis, cough, chest pain, hemoptysis, or contact with any smear-positive pulmonary tuberculosis.

S. Jahan, *Endocrinology Case Studies*,
https://doi.org/10.1007/978-981-95-2770-0_20

Fig. 20.1 Ill-looking and emaciated appearance

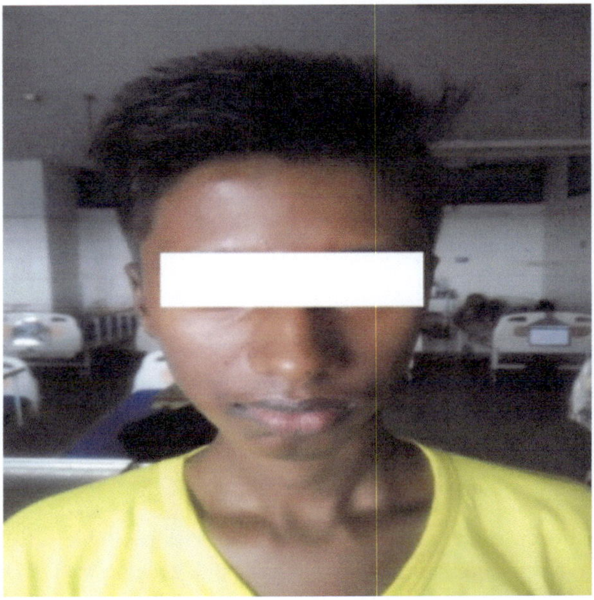

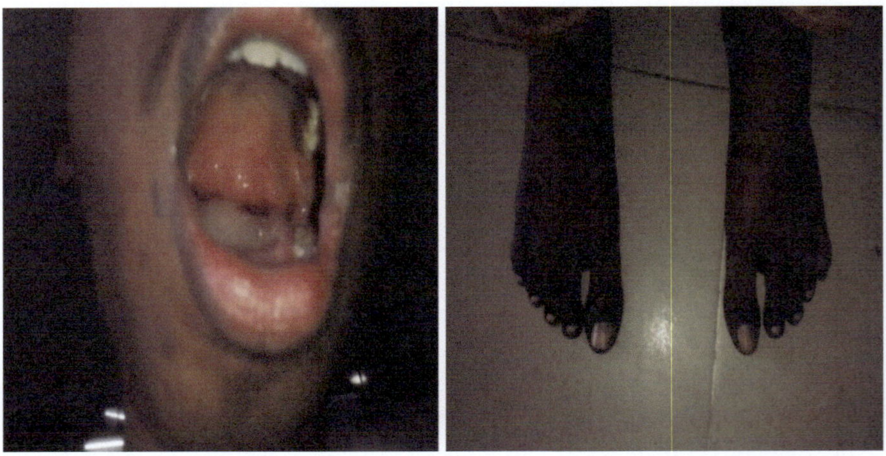

Fig. 20.2 Hyperpigmentation of oral mucosa and extremities

On examination, he was ill-looking and emaciated and mildly anemic; generalized pigmentation was present more on lips, hands, and feet (Figs. 20.1, 20.2 and 20.3). His height was 157.5 cm, weight: 53 kg, and BMI: 21.54 kg/m^2. His blood pressure was 110/80 mmHg, no postural drop, and pulse rate was 80 beats/min. Neurological examination showed absence of vibration and joint position sense up to ankle joint with positive Romberg sign; the plantar response was flexor; he had normal motor function, and cranial nerve examination, including funduscopy, was unremarkable. Other systemic examinations including chest and abdomen revealed no abnormality.

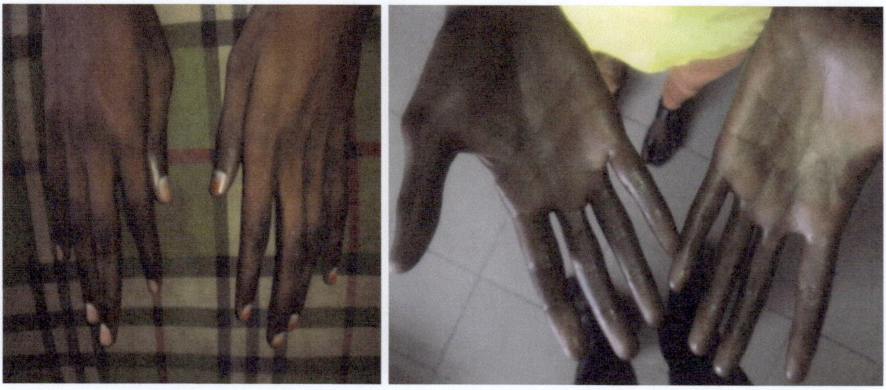

Fig. 20.3 Hyperpigmentation of palms of both hands

Q1. What is the probable differential diagnosis depending on clinical features?
  (a) Addison disease
  (b) APS type-1
  (c) Kala-azar
  (d) Tuberculosis (TB)
  (e) Pernicious anemia

All of the above diagnoses are possible in this boy. He had weight loss, hyperpigmentation, and weakness suggestive of Addison disease, though the absence of hypotension or H/O any adrenal crisis and pigmentation of classic sites like hard palate and nail plate disfavored it. On the other hand, APS type-1 consists of Addison disease along with tetany resulting from hypoparathyroidism, mucocutaneous candidiasis, and ectodermal dystrophy, all of which were absent in this boy. He did not come from an endemic area of Kala-azar, and the absence of significant anemia, documented fever, and splenomegaly made this diagnosis less likely. He did not have H/O contact with smear-positive TB, cough, chest pain, hemoptysis, or any evening rise of temperature, and his respiratory system examination failed to reveal any abnormality. Thereby, TB was also an unlikely diagnosis here. Pernicious anemia was possible considering the combination of moderate anemia, pigmentation, and neurological deficit suggestive of subacute combined degeneration (SCD) of the spinal cord. However, his normal diet pattern and bowel habit, absence of other autoimmune diseases, and symptoms/signs suggestive of peripheral neuropathy or pyramidal tract involvement were points to be considered against this diagnosis.

The routine investigations showed mild macrocytic anemia with iron deficiency, and the presence of UTI, serum electrolyte, and calcium panel was normal. Both morning and evening cortisols were inappropriately low (Table 20.1).

Q2. What other investigations are required at this point?
  (a) Short synacten test (SST)
  (b) Repeat basal plasma cortisol

Table 20.1 Hematologic and biochemical parameters of the patient showing macrocytic anemia and subnormal plasma cortisol

Investigations	Result	Reference range
CBC	Hb%—**11.52gm/dL**, ESR—24 mm, MCV—**96 fl** WBC—8100/mm^3, DC-N—60%, L—33%, E—05% M—02% PBF: Nonspecific findings	Hb—15 ± 2 g/dL, ESR—0–10 mm in first hour, TC of WBC—7 ± 3 × 10^9/L
S. Bilirubin	0.8 mg/dL	0.2–1 mg/dL
SGPT	22 U/L	14–63 U/L
Serum iron	38 ug/dL	60–170 ug/dL
Serum TIBC	225 ug/dL	250–450 ug/dL
Serum ferritin	85.2 ug/dL	20–350 ug/dL
Urine R/M/E	**Pus cell—5–8/HPF**	–
Serum calcium	9.6 mg/dL	8.5–10.5 mg/dL
Serum electrolyte	Na-139 mmol/L, K—4.6 mmol/L, Cl—103 mmol/L, TCO—22.7 mmol/L	Na—135–144, K—3.5–5.2, Cl—95–107, TCO_2—22–30 mmol/L
RBS	7.3 mmol/L	–
Hb electrophoresis	Normal study	–
Basal plasma cortisol	**5.90 ug/dL**	1.6–19.40 ug/dL
Evening cortisol	**3.60 ug/dL**	2.90–17.30 ug/dL

(c) Serum /RBC folate level
(d) UGIT endoscopy
(e) Serum Vit B12 level

Basal plasma cortisol is rarely diagnostic of Addison disease because of marked variation related to various factors including the timing of the day, stress, intercurrent illness, etc. Basal cortisol below 50 nmol/L is likely to exclude hypercortisolism, whereas level > 550 nmol/L excludes hypocortisolism essentially. However, none of the values are sensitive enough for a concrete diagnosis. On the other hand, hormonal tests if not done properly with sensitive assays like third generation immunoassays or HPLC/TMS are prone to false results. So, it is better to repeat them before making a conclusion. SST is the confirmatory test for Addison disease that is mandatory for this case. Low serum Vit B 12 level is suggestive of pernicious anemia that might occur due to the presence of anti-intrinsic factor antibody. In that case, other autoimmune diseases might be present as well. UGIT endoscopy is indicated to evaluate gastric pathology in such a case. Folate deficiency leads to macrocytic anemia but does not cause any neurologic problems. So, serum/RBC folate is not necessary here.

Table 20.2 Specific investigation report of the patient showing normal HPA axis, low serum vit-B12, and gastritis on UGIT endoscopy

Investigations	Result	Reference range
FT4	1.2 ng/dL	0.8–1.9 ng/dL
TSH	3.5uIU/mL	0.3–4.5uIU/mL
ACTH	63.7 pg/mL	ND to 46 pg/mL
B. Cortisol	404.3 nmol/L	138- 690 nmol/L
After SST, S. Cortisol	1015.76 nmol/L	>550 nmol/L
Serum PTH	46 ng/dL	14–65 pg/mL
Vitamin B12	**164 pg/mL**	239–931 pg/mL
ICT for kalazar	IgG negative IgM negative	–
ANA	Negative	–
Monteoux test	Negative	–
Upper GI endoscopy	**Multiple erosions seen at body of stomach** **Impression—Gastritis**	–
Anti-IF antibody	**Positive**	<1.2 AU/mL—Negative >1.53 AU/mL—Positive

Specific investigations were done subsequently to establish the diagnosis.

On repeat testing, basal and stimulated cortisols following SST were found to be normal. Plasma ACTH was also normal, thereby excluding Addison disease. TFT and serum PTH were normal enough to rule out APS type-1. Serologic tests including ICT for Kala-azar proved negative, and tests to exclude TB like MT and CXR were also unremarkable. Thereby, infectious diseases on our list were excluded effectively. The patient had significantly low serum vit-B12, establishing SCD due to its deficiency. To find out the underlying cause, UGIT endoscopy was done that showed gastritis. Autoimmune screening for pernicious anemia was positive, though ANA was found to be negative (Table 20.2). He was started on vit B12 injection as per the recommended schedule and showed significant improvement on subsequent follow-up.

Q3. The factors that indicate successful treatment with vit B12 are:
  (a) Rising reticulocyte count
  (b) Increasing hemoglobin level
  (c) Decrease in serum K level
  (d) Improvement of sensory impairment
  (e) Increase in serum K level

Successful treatment is indicated by reticulocytosis on the 5th–tenth day, rise of Hb of 1 gm/dL every week, a decrease of plasma K, and depletion of the iron store. This patient also had a concurrent iron deficiency that required correction as well. In general, treatment with vit-B12 leads to depletion of the iron store. The neurologic features in pernicious anemia might be irreversible.

Q4. What factors need to be monitored during long-term follow-up?
   (a) FT4, TSH
   (b) CBC, ESR with PBF
   (c) Basal cortisol, plasma ACTH
   (d) ANA
   (e) Serum holo-trans-cobalamin level

Pernicious anemia is an autoimmune disease. While looking for clinical and biochemical improvements in terms of correction of anemia and other symptoms, vigilant monitoring for other autoimmune conditions—including Graves' disease, Hashimoto's thyroiditis, Addison disease, and SLE—is warranted considering the clinical context. So, all the above answers are true except E. Reduction in serum holo-transcobalamin level is a surrogate marker for the diagnosis of pernicious anemia. It is found to be low in such a case. However, it need not be monitored during regular follow-up.

Discussion

Fever and weight loss with hyperpigmentation can be the protean manifestation of many systemic diseases like TB, Kala-azar, Addison disease, or pernicious anemia. Hyperpigmentation in the backdrop of macrocytic anemia and neurologic features should be evaluated for ***pernicious anemia***. Pernicious anemia (PA) is macrocytic anemia that is caused by vitamin B12 deficiency because of intrinsic factor deficiency. PA is associated with atrophic body gastritis, where the diagnosis is based on histological confirmation of gastric body atrophy [1]. Though the term PA is sometimes used synonymously with cobalamin deficiency, to avoid ambiguity, the term is commonly preserved for conditions that result from impaired secretion of intrinsic factors and atrophy of the oxyntic mucosa [2, 3]. It is mainly considered a disease of the elderly, but younger patients represent about 15% of patients [1]. It is frequently associated with autoimmune thyroid disease (40%) and other autoimmune disorders, such as diabetes mellitus (10%), as part of the autoimmune polyendocrine syndrome [1]. It is recognized that most of these patients present mainly with neurologic symptoms [3]. However, hyperpigmentation could be the earliest manifestation of vitamin B12 deficiency even before anemia sets in. Anemia occurs in a patient with severe vitamin B12 deficiency for a longer duration. In our patient, the hemoglobin level was only mildly diminished for age and sex. He presented with hyperpigmentation and neurologic symptoms. The cutaneous manifestation of vitamin B12 deficiency is skin hyperpigmentation, vitiligo, hair changes, and recurrent angular stomatitis. Hyperpigmentation of the extremities, especially over the dorsum of the hands and feet, with accentuation over the interphalangeal joints and terminal phalanges associated with pigmentation of the oral mucosa is characteristic of vitamin B12 deficiency [4, 5]. Diagnosis can be made with serum B12 < 350 pg/mL (90% sensitive), anti-intrinsic factor antibody (50% sensitive, 100% specific), and antiparietal cell antibody (80% sensitive, 50–100% specific). The treatment of vitamin B12 deficiency due to malabsorption/PA is parenteral hydroxocobalamin and is lifelong, with the recurrence of symptoms on discontinuation.

Answer

Q1. (a), (e)
Q2. (a), (b), (d), (e)
Q3. (a), (b), (c)
Q4. (a), (b), (c), (d)

References

1. Lahner E, Annibale B. Pernicious anemia: new insights from a gastroenterological point of view. World J Gastroenterol. 2009;15:5121.
2. Babior BM. Erythrocyte disorders. Anemias related to disturbance of DNA synthesis (megaloblastica anemias). In: Williams JW, Beutler E, Erslev AJ, Lichtman MA, editors. Hematology. 4th ed. New York: McGraw-Hill; 1998. p. 453–81.
3. Lindenbaum J, Healton EB, Savage DG, Brust JC, Garrett TJ, Podell ER, Marcell PD, Stabler SP, Allen RH. Neuropsychiatric disorders caused by cobalamin deficiency in the absence of anemia or macrocytosis. N Engl J Med. 1988;318(26):1720–8.
4. Baker SJ, Ignatius M, Johnson S, Vaish SK, Hyperpigmentation of skin. A sign of vitamin-B12 deficiency. Br Med J. 1963;1:1713–5.
5. Ridley CM. Pigmentation of fingertips and nails in vitamin B12 deficiency. Br J Dermatol. 1977;97:105–6.

A 25-Year-Old Male with Suspected Adrenal Crisis and Hyperpigmentation

21

Introduction

The following case scenario exemplifies the intriguing manifestation of pheochromocytoma, making it a difficult puzzle to solve. A young farmer, presumed to have adrenal crisis based on hypotensive shock, hyperkalemia, and AKI, was started on steroid replacement therapy. The diagnosis was questioned as he admitted of repeated panic spell, hypertensive retinopathy, and fluctuating BP. Subsequently, his investigations revealed the adequacy of HPA axis and bilateral adrenal tumor that proved to be pheochromocytoma on histopathologic examination. His age and mode of presentation pointed toward a genetic syndrome, possibly VHL, and he was investigated accordingly.

Case Scenario

A 25-year-old farmer presented with repeated episodes of vomiting and headache for the last 5 years. He was hospitalized following an episode of vomiting, profuse sweating, and palpitations 5 years back at a local hospital. At that time, he was found to have a nonrecordable BP and managed with IV fluids and pressor drugs but then discharged without a proper diagnosis. He remained free of symptoms for the next 2 years. Then, the second episode occurred that necessitated similar treatment at another hospital. One year later, he presented with fever and severe upper abdominal pain with vomiting and was diagnosed as having cholelithiasis. Cholecystectomy was planned which was, however, postponed due to a sudden surge in BP during the anesthetic procedure. It was successfully performed later on under antihypertensive coverage which was discontinued shortly afterward. Two years later, he had several episodes of headache, vomiting, and sweating over 1 week that compelled his

S. Jahan, *Endocrinology Case Studies*,
https://doi.org/10.1007/978-981-95-2770-0_21

admission into a tertiary care hospital. His pulse and BP were nonrecordable, and he was diagnosed as having pre-renal AKI and hyperkalemia with adrenal crisis (serum creatinine, 3.2 mg/dL; serum potassium, 7.3 mmol/L). He was managed with IV fluids and injectable hydrocortisone. Then, he was discharged home on the steroid without undergoing any definitive investigations to confirm hypoadrenalism. He continued to have recurrent spell of headache and vomiting that compelled him to consult us.

On examination, he was a depressed yet cooperative gentleman having a BMI of 18.73 kg/m^2. He had generalized hyperpigmentation, with an incision mark over the right hypochondriac region. His pulse rate was 78 b/min, regular, and BP was 110/70 mm Hg (no postural drop). The systemic examination was unrevealing except for the presence of grade IV hypertensive retinopathy on fundoscopy.

Q1. What were the points in history suggesting an adrenal crisis?
- (a) Shock/hypotension
- (b) Acute abdominal pain
- (c) Hyperpigmentation
- (d) Hyperkalemia/azotemia
- (e) Unexplained fever

Adrenal crises may sometimes present with unexplained low-grade fever, vomiting, or an acute abdomen. A hyperpigmented patient is often found to be in a state of hypotensive shock/nonrecordable BP, while depletional hyponatremia with hyperkalemia and pre-renal azotemia is a common finding in such a scenario. However, the diagnosis was not confirmed with necessary investigations like basal cortisol, plasma ACTH, or short synacthen test. There were some unusual features like episodes of headache, palpitations, and sweating and the evidence of grade IV hypertensive retinopathy that pointed toward long-standing hypertension rather than hypotension in this patient. So, we started to record the BP at regular intervals daily, and the chart is shown in Table 21.1.

Table 21.1 Chart of ambulatory BP monitoring of the patient showing fluctuating HTN

Date	Time	BP (mm of Hg)
23.2.16	8.30 am	**140/110**
	12.50 pm	**180/140**
	5 pm	**120/100**
24.2.16	2 pm	100/80
	5 pm	110/80
25.2.16	12 pm	**180/130**
3.3.16	8.30 am	120/80
6.3.16	8 am	**140/110**
	2 pm	**160/120**

Q2. What might be the most probable diagnosis for him?
(a) Renovascular hypertension
(b) Labile essential hypertension
(c) Pheochromocytoma
(d) Intracranial space-occupying lesion (ICSOL)
(e) Migraine

The patient had fluctuating hypertension on ambulatory BP monitoring (Table 21.1). Now, we had to rethink the diagnosis. Steroid preparation was stopped following biochemical evaluation showing the adequacy of the HPA axis. All the above diagnoses seemed possible in the context of the current scenario. The pretest probability for pheochromocytoma was high considering the spells of headache, vomiting, and abdominal pain. Probably, these all accounted for the classic Pheo spell. As there was a history of AKI with hyperkalemia, renovascular hypertension was also possible, but there was no renal bruit or any sign of atherosclerotic disease elsewhere or evidence of renal damage like proteinuria, etc., and on repeat testing, the serum creatinine and electrolyte became normal. So, this was an unlikely diagnosis here. He had headache and vomiting with hypertension, making ICSOL a possibility. For the same reason, we also had to consider migraine headache. Lastly, labile essential hypertension was in our mind too. However, all the features could not be explained by either of them.

After scrutinizing history, certain important facts became evident. The hyperpigmentation in this patient was probably occupational, resulting from chronic sun exposure while farming. He had a repeated Pheo spell that ended with profuse vomiting and sweating requiring hospitalization. Hypotension at that time could be attributed to severe vomiting, chronic volume depletion, and impaired sympathetic reflexes, all of which are common in pheochromocytoma. The patients with pheochromocytoma often tend to have considerable natriuresis at the end of an episode that may also lead to a shock-like state or hypotension. Interestingly there was a history of pressor response during anesthesia for cholecystectomy that provided another clue to the diagnosis.

The patient had polycythemia, erythrocytosis, and hyperglycemia, all of which can sometimes occur as paraneoplastic phenomena in pheochromocytoma (Table 21.2). The low normal cortisol and high ACTH indicated the recovery of the HPA axis from steroid-induced suppression, which was later confirmed with SST. However, 24 h urinary metanephrine was found to be normal unexpectedly (Table 21.3). As the pretest probability of pheochromocytoma was considerably high in this patient, we advised him to undergo an adrenal CT scan anyway. It showed well-defined, sharply marginated soft tissue masses with a solid and intralesional small cystic component having a lipid attenuation value of 45–56 HU (precontrast) over both the adrenal regions (Fig. 21.1). Interestingly, USG of abdomen done previously was completely unremarkable.

Table 21.2 Biochemical report of the patient showing polycythemia, hyperglycemia, and urinary abnormalities

Investigation	Result	Reference value
CBC	Hb—**17.8 g/dL**, RBC—**824,000**/mm^3 TC of WBC—8500/mm^3 Platelet count—330,000/mm^3	Hb—15 ± 2 g/dL, RBC—5 ± 0.5 × 10^{12}/l TC of WBC—7 ± 3 × $10^{9/l}$ Platelet— 150,000–400,000/L
Serum creatinine	0.8 mg/dL	0.5–1.3 mg/dL
Serum electrolyte	Na—142 mmol/L, K—3.7 mmol/L, Cl—102 mmol/L, TCO_2—24.8 mmol/L	Na—135–145, K—3.5–5.1, Cl—95–107, TCO2—22–30 mmol/L
Blood sugar (OGTT)	FBS—6.8 mmol/L 2 h OGTT—**14.8 mmol/L**	Normal: FBS ≤ 5.5 mmol/L, 2 h OGTT ≤7.7
USG of whole abdomen	Bilateral early renal parenchymal change along with cystitis	–
ECG	Sinus tachycardia	–
Urine R /E	**Protein—+** **Sugar—+++**	–

Table 21.3 Hormonal report of the patients showing normal HPA axis and elevated urinary nor-metanephrine

Name	Result	Reference value
FT4	1.35 ng/dL	0.8–1.8 ng/dL
TSH	1.86 μU/mL	0.4–4.5 mU/L
Serum cortisol	**117.8 nmol/L**	**138–579.4 nmol/L**
ACTH	**80.2 pg/mL**	**Up to 46 pg/mL**
Short synacthen test (plasma cortisol)	**670 nmol/L**	**Adequate HPA axis > 550 nmol/L**
DHEAS	**221 mcg/dL**	**38–313 mcg/dL**
24 h urinary metanephrine	**690 nmol/day**	**126.8–1125.6 nmol/day**
24-hr-urinary normetanephrine	**967 mcg/day**	**75–375 mcg/day**
Serum calcium	8.9 mg/dL	8.5–10.5 mg/dL
PTH	34 pg/mL	14–65 pg/mL
Serum calcitonin	15 pg/mL	≤19 pg/mL

The patient was managed with multiple antihypertensives including prazosin, amlodipine, beta-blockers, and spironolactone. Subsequently, he had bilateral adrenalectomy without any intra- or postoperative complications. Histopathology of the masses revealed bilateral benign pheochromocytoma with typical Zell Ballen appearance with positive immunostaining for neuroendocrine tumor. On subsequent follow-up, he was found to be normotensive and normoglycemic without any drugs and supplemented lifelong with tab hydrocortisone.

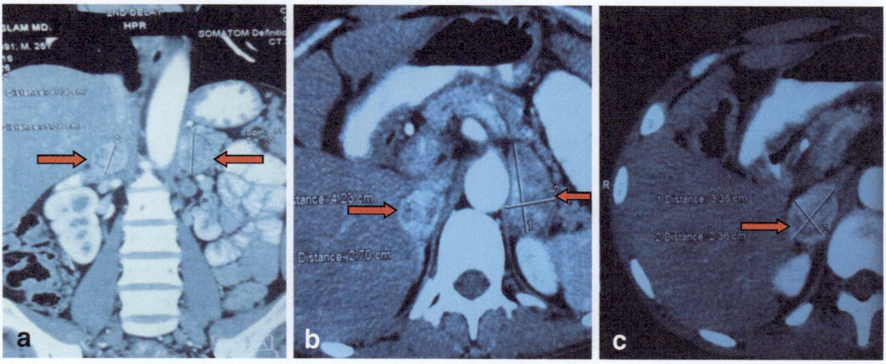

Fig. 21.1 (**a**–**c**) Bilateral adrenal masses suggestive of pheochromocytoma

Q3. What are the features in this patient necessitating genetic testing for pheochromocytoma?
(a) Age < 30 years
(b) Bilateral pheochromocytoma
(c) The normal level of urinary metanephrine
(d) Co-phenotype disorder
(e) Fluctuating HTN

This young hypertensive male without any significant family history or co-phenotype disorder presented with bilateral pheochromocytoma. Certainly, his age and bilaterality of the tumor were strong pointers to a familial syndrome associated with pheochromocytoma. MEN 2A/2B, Von Hippel–Lindau syndrome (VHL) commonly, NF1, Sturge Weber syndrome, tuberous sclerosis, MEN 1, or Carny's complex in some cases are associated with pheochromocytoma. The presence of positive family history and co-phenotype disorder like MTC and hyperparathyroidism in MEN 2A or MTC and mucosal neuroma/marfanoid habitus in MEN 2B are suggestive of the underlying syndrome. Our patient had a normal level of serum calcitonin and calcium and an unremarkable neck USG, thereby excluding MEN 2 syndrome. VHL is characterized by cerebellar and spinal hemangioblastoma, retinal angioma, renal cell carcinoma, etc., whereas NF1 is associated with a cutaneous neurofibroma, optic nerve glioma, skeletal abnormality, etc. There are also some specific dermatologic and musculoskeletal manifestations of these genetic disorders that might provide important diagnostic clues like lichen amyloidosis, angioma, collagenoma, lipoma, etc. Sometimes, the nature of the plasma/urinary metabolite might point toward the underlying syndrome, for example, elevated epinephrine/metanephrine in MEN syndrome and nor-epinephrine/nor-metanephrine in VHL. Also, the location and laterality are important in deciding the type of genetic testing: bilateral pheochromocytoma in MEN 2A/B and VHL, unilateral and apparent sporadic pheochromocytoma in NF1, and paraganglioma in SDH B, C, D mutation. Fluctuating HTN can occur in 30% of pheochromocytoma cases, and it is not specifically related to any genetic syndrome. It is related to the episodic release of catecholamine from the tumor.

Our patient had a normal level of metanephrine, calcitonin, and calcium, thereby having a low probability of MEN syndrome. He was tested for 24 h urinary normetanephrine before surgery which was found to be high. So, he possibly had VHL, and we are planning to exclude the other co-phenotype disorder and undertake genetic testing, followed by his family screening in the future.

Discussion

Pheochromocytoma commonly arises as a unilateral adrenal mass, either symptomatic or asymptomatic. In 10% of cases, especially in ***PPGL syndrome***, the tumors are bilateral. In the syndromes of multiple endocrine neoplasia (MEN) type 2a and type 2b and, in certain families with ***Von Hippel–Lindau (Lindau's) disease*** [1, 2], bilateral pheochromocytomas are the rule. In neurofibromatosis, pheochromocytomas are an uncommon manifestation but may affect both adrenal glands in the exceptional case [3]. In most of these cases of ***bilateral pheochromocytoma,*** genetic mutation has already been identified. Yet in a few cases, they may be sporadic and not manifesting any specific clinical, biochemical, or genetic phenotype. Identification of new hereditary forms of PPGL has led to the highest rate of germline susceptibility in cancer genetics at almost 40% [4, 5]. In addition, other PPGL-related genes have been discovered, and there are currently over 22 susceptibility genes identified. The genotype–phenotype correlation shown in many studies often dictates the clinical presentation of syndromic forms of the disease. These include associated biochemical profile, tumor location, malignant potential, aggressive clinical behavior, and overall prognosis. Hereditary catecholamine-secreting tumors typically manifest at a younger age than sporadic neoplasms do. Genetic testing should be considered if a patient has one or more of the following: (a) paraganglioma, (b) bilateral adrenal pheochromocytoma, (c) unilateral adrenal pheochromocytoma and a family history of pheochromocytoma/paraganglioma, (d) unilateral adrenal pheochromocytoma with onset at a young age (<30 years), or (5) other clinical findings suggestive of one of the previously discussed syndromic disorders [6]. The patient in the case scenario was only 25 years of age presented with fluctuating HTN and panic attacks. He had an elevated nor-metanephrine and bilateral adrenal tumor. Though there was no apparent abnormality of any co-phenotype disorder, he was likely to have one with the strongest possibility of VHL.VHL is an autosomal dominant disorder characterized by a wide variety of benign and malignant neoplasms. It is caused by the VHL tumor suppressor gene, located on chromosome 3p25–26, which encodes a protein that regulates hypoxia-induced proteins [7]. There is a strong genotype-to-phenotype correlation and nearly 100% penetrance. The hypoxia-induced protein leads to the stimulated synthesis of erythropoietin which in turn results in polycythemia and erythrocytosis as in this case. The initial screening tests for VHL include measurement of metanephrines, MRI of the brain, spinal cord, and abdomen, and performing ophthalmoscopy. Following surgery, metanephrines should be measured yearly, and an MRI of the brain, spinal cord, and abdomen along with ophthalmoscopy should be done also at an annual interval; if no tumor is found, monitoring is to be continued every 2 or 3 years [8].

Answer

Q1. all
Q2. (c)
Q3. (a), (b)
Q4. (a), (b)

References

1. Lips KJM, Van Der Sluys Veer J, Struyvenberg A, Alleman A, Leo JR, Wittebol P, Minder WH, Kooiker CJ, Geerdink RA, Van Waes PFGM, Mackeng WHL. Bilateral occurrence of pheochromocytoma in patients with multiple endocrine neoplasia type 2a (Sipple's syndrome). Am J Med. 1981;70:1051.
2. Atuk ND, McDonald T, Wood T, Carpenter JT, Walzak MP, Donaldson M, Gillenwater JY, Turner SM, Westfall V. Familial pheochromocytoma, hypercalcemia, and Von Hippel-Lindau disease. Medicine. 1979;58:209.
3. Kalff V, Shapiro B, Lloyd R, Sisson JC, Holland K, Nakajo M, Beierwaltes WM. The spectrum of pheochromocytoma in hypertensive patients with neurofibromatosis. Arch Intern Med. 1982;152:2092.
4. Castro-Vega LJ, Letouzé E, Burnichon N, Buffet A, Disderot PH, Khalifa E, et al. Multi-omics analysis defines core genomic alterations in pheochromocytomas and paragangliomas. Nat Commun. 2015;6:6044. https://doi.org/10.1038/ncomms7044.
5. Jochmanova I, Pacak K. Genomic landscape of pheochromocytoma and paraganglioma. Trends Cancer. 2018;4:6–9. https://doi.org/10.1016/j.trecan.2017.11.001.
6. William F Young Jr .Endocrine HTN. William's textbook of Endocrinology, 12th ed.
7. Hes FJ, Hoppener JW, Lips CJ. Clinical review 155: pheochromocytoma in Von Hippel-Lindau disease. J Clin Endocrinol Metab. 2003;88:969–74.
8. Neumann HPH, Young WF Jr, Eng C. Pheochromocytoma and paraganglioma. N Engl J Med. 2019;381:6. nejm.org.

22 A 21-Year-Old Male with Cushing Syndrome Following Pituitary Surgery for Nonfunctioning Adenoma

Introduction

A young male having a history of TSA for presumed nonfunctioning pituitary adenoma developed clinical and biochemical features of Cushing syndrome starting a year after the surgery. The etiology of Cushing was uncertain; it was initially suspected to be drug-induced but later excluded based on the history and elevated plasma ACTH and cortisol levels. Ectopic Cushing syndrome could not be confidently ruled out, and repeat pituitary surgery was undertaken because of the recurrence of NFA with optic chiasmal compression. Finally, the positive immunostaining for ACTH in the resected tumor established the diagnosis of CD where a cotico-troph adenoma that was previously silent had become functionally active producing the stigmata of Cushing syndrome.

Case Scenario

A 21-year-old nonalcoholic, nonsmoker male presented with progressive weight gain over the past 2 years. Three years back, he was evaluated for a gradual visual disturbance with narrowing of the visual field and diagnosed to have pituitary macroadenoma (Figs. 22.1 and 22.2). It was not associated with headache, nausea, vomiting, or diplopia, and there were no clinical features suggestive of hormonal hypersecretion or hypopituitarism. Hormonal investigations were unremarkable at that time (Table 22.1), and he was diagnosed as a case of nonfunctioning pituitary macroadenoma (NFA). Following that, he underwent trans-sphenoidal adenomectomy (TSA), and a postsurgical histopathology report confirmed pituitary adenoma. However, an immunohistochemical study was not advised in that instance. During the next 3 weeks in the perioperative period, he received oral dexamethasone. His visual disturbance did not completely improve, so he took various types of steroid-containing medications on and off for the next 1 year. Soon after the operation, he

S. Jahan, *Endocrinology Case Studies*,
https://doi.org/10.1007/978-981-95-2770-0_22

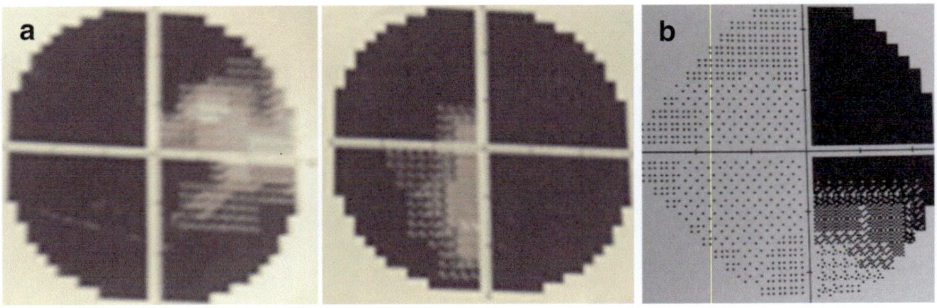

Fig. 22.1 Perimetric changes (**a**) before first operation (bilateral, almost complete anopia); (**b**) after second operation (bilateral, superotemporal quadrantanopia with some involvement of inferotemporal quadrants)

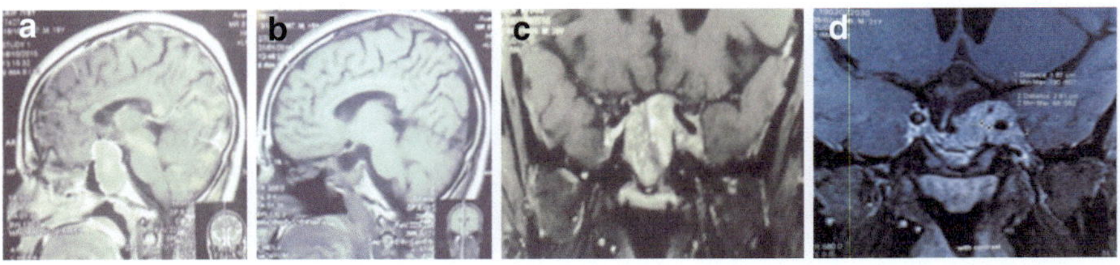

Fig. 22.2 MRI of pituitary macroadenoma in relation with surgery (**a**) before first operation (sellar mass with suprasellar extension); (**b**) after first operation (residual sellar mass with postoperative fluid collection); (**c**) before second operation (recurrence of the tumor); (**d**) after second operation (partial empty sella with residual mass)

Table 22.1 Initial hormonal investigation results within normal limits

Tests	Result	Reference value
S. TSH	2.21 μIU/mL	0.35–5.5 μIU/mL
S. FT4	1.27 ng/dL	0.8–1.8 ng/dL
S. Cortisol	15.34 μg/dL	5–25 μg/dL
S. Prolactin	6.4 ng/mL	2.2–8.5 ng/mL
Plasma ACTH	34 ng/mL	10–65 ng/mL

noticed progressive weight gain and the appearance of purplish striae over the abdomen. As his visual complaint remained unchanged, he stopped these herbal medications abruptly altogether. At that time, he did not develop any adrenal crisis, and his weight remained static and the striae turned white. However, over the last 2 years, he had not been on any steroid-containing medication. Even then, he started gaining considerable weight with the reappearance of purple striae and gradual deterioration of his vision. These complaints led to his admission into our hospital.

He was an obese gentleman (BMI, 33 kg/m^2), having a moon face with a plethora. His BP was 140/100 mm Hg without any postural drop. There was prominent central obesity, acanthosis nigricans, dorsocervical fat pad, and supraclavicular

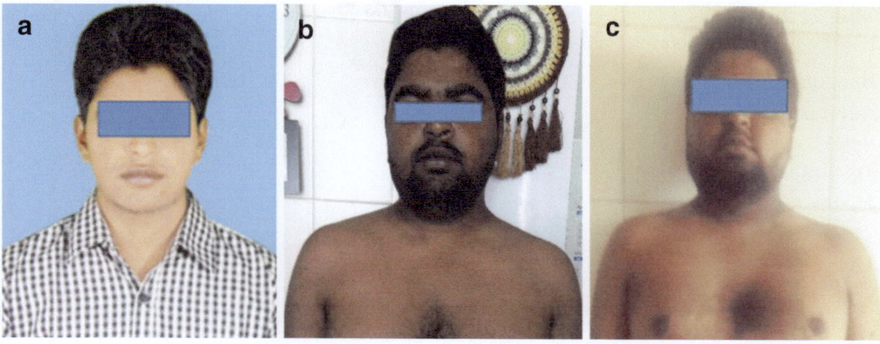

Fig. 22.3 Facial appearance of the patient. (**a**) Before first surgery (Normal); (**b**) before second surgery (Cushingoid); (**c**) after second surgery (regression of Cushingoid appearance)

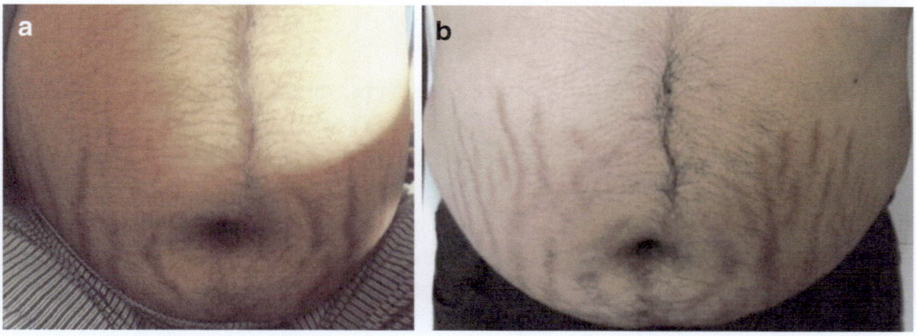

Fig. 22.4 Abdominal striae (**a**) before second operation (active), (**b**) after second operation (in regression)

fullness (Fig. 22.3). There were wide, depressed red–purple striae over the lower abdomen and widespread fungal infection in extremities (Fig. 22.4). Ophthalmologic examination showed diminished visual acuity (right eye—6/12, left eye—counting of finger) and global restriction of the visual field on confrontation, and fundoscopy showed temporal pallor in the right disc as well as pallor in the left disc with normal-appearing macula.

Q1. What are the possibilities at this moment?
- (a) Iatrogenic Cushing syndrome
- (b) Cushing disease
- (c) Pseudo-Cushing syndrome
- (d) NFA (recurrence/persistence)
- (e) None of the above

Iatrogenic Cushing's syndrome was a strong possibility as there was significant drug history and clinical and biochemical evidence of hypercortisolism. However,

Table 22.2 Hormonal investigations during subsequent presentation with Cushing's syndrome showing high plasma cortisol and ACTH

Hormone profile	Result	Reference value
Adrenal axis		
Basal cortisol (30.7.18)	**1328 nmol/l**	138–690 nmol/L
(08.8.18)	**923.2 nmol/L**	
Plasma ACTH	**173 pg/mlml**	Not detectable–46 pg/mL
	153.8 pg/mL	
Thyroid axis		
S. TSH	1.29 uIU/mL	0.35–5.5 μIU/mL
S. FT4	1.25 ng/dL	0.8–1.8 ng/dL
Gonadal axis		
S. LH	3.93 mIU/mL	0.8–7.8 mIU/mL
S. FSH	**1.80 mIU/mL**	0.7–11.1 mIU/mL
S. Testosterone	**152.5 ng/dL**	270–1734 ng/dL
S. Prolactin	3.10 ng/mL	2.2–8.5 ng/mL

he continued to have features of CS long after the cessation of steroid intake and did not have any withdrawal symptoms or adrenal crisis after that. The initial biochemical reports were less reliable as samples were collected on an OPD basis while the patient was receiving itraconazole. Obesity with depressive illness arising out of chronic comorbidities coupled with blindness could be a cause of the pseudo-Cushing syndrome. The absence of significant catabolic features like skin thinning, easy bruising, proximal myopathy, and hypokalemia was suggestive, and on the other hand, lack of suppression of plasma cortisol on LDDST made it less likely. The patient had both metabolic and catabolic features of Cushing's syndrome and no features of adrenal insufficiency following steroid discontinuation with very high basal plasma cortisol and ACTH (Table 22.2). So, Cushing disease was the strongest possibility. He had a residual tumor with compressive symptoms that could explain a few of his presentations, but CS would not be explained by the recurrence of NFA. Initially, we were in favor of nonfunctioning pituitary macroadenoma (post-surgical state) with drug-induced Cushing's syndrome.

Routine hematologic and biochemical investigations showed neutrophilic leukocytosis, elevated liver enzyme, impaired fasting and post-glucose values, and evidence of combined dyslipidemia. Other tests including serum creatinine, BUN, and electrolytes (Na—145, K—4.5, Cl—104, HCO3—27.4 mmol/L) were all within normal limits. Then, we embarked on the hormonal investigations. He had low serum prolactin and hypogonadotropic hypogonadism with the normal thyroidal axis. Plasma basal cortisol and ACTH were found to be significantly high on several occasions; 24 h urinary free cortisol was moderately elevated twice, and normal once, on repeated testing (Tables 22.2 and 22.3).

So, to confirm Cushing's syndrome, we performed an overnight low-dose dexamethasone suppression test (LDDST) which was also done previously at another center. The plasma cortisol remained unsuppressed both the times, confirming Cushing syndrome biochemically. Drug-induced Cushing was virtually excluded

Table 22.3 Positive screening for Cushing's syndrome

Tests	Result	Reference value
24 h urinary free cortisol		
(9. 8.18)	**242 ugm/24 h**	50–190 μg/24 h
(11.8.18)	**267 ugm/24 h**	
S. Cortisol after ONLDDST (12. 8.18)	**686.7 nmol/L**	<50 nmol/L

ONLDDST Overnight low-dose dexamethasone suppression test; Date (D/M/Y) within parentheses

Table 22.4 Nonsuppressed plasma cortisol following overnight high-dose dexamethasone suppression test

Test	Result	Reference value
Basal cortisol	**1328 nmol/L**	138–690 nmol/L
S. Cortisol after ONHDDST	**847.5 nmol/L**	Nonsuppressed: <50% decremental value

ONHDDST Overnight high-dose dexamethasone suppression test

because he was off the offending drugs for the last 2 years, and both plasma cortisol and ACTH were found to be high rather than low. So, he represented a case of endogenous Cushing syndrome at this point (Table 22.3).

To find out the etiology, we repeated plasma ACTH, and it was reasonably high, indicating an ACTH-dependent Cushing syndrome. Plasma cortisol remained >90% of baseline level following the overnight high-dose dexamethasone suppression test (HDDST), which raised the possibility of ectopic Cushing syndrome (Table 22.4). For localization of the tumor, he underwent HRCT of the chest and abdomen, and both were unremarkable.

Q2. What is the probable diagnosis at this point?
  (a) Ectopic Cushing syndrome
  (b) Cyclic Cushing syndrome
  (c) Silent corticotroph adenoma
  (d) Cushing disease arising de novo on NFA
  (e) None of the above

Cyclic Cushing syndrome is characterized by periods of excess cortisol interspersed with intervals of normal cortisol production. Some of the patients demonstrate a paradoxical rise in plasma ACTH and cortisol when treated with dexamethasone, and occasionally a patient is benefitted from dopamine agonist (bromocriptine or cabergoline) therapy. Most patients have pituitary-dependent disease, and in many of these patients, basophil adenomas have been removed, with a long-term cure in some cases. Our patient had progressive deterioration of Cushing syndrome with persistently high plasma ACTH and cortisol, making cyclic Cushing an unlikely diagnosis. Ectopic Cushing syndrome may be caused by small NET, and in about 12–13% of cases, no tumor could be identified even after extensive search. The coexistence of ectopic Cushing and pituitary NFA at the same time looks improbable though not impossible, and it is always wise to look for a single diagnosis. For the same reason, the recurrent NFA with de novo Cushing disease was also less likely. Then, we reviewed our provisional diagnosis, and Cushing disease now seemed a possibility.

Q3. What should be a judicious decision regarding the reports?
  (a) Refer to a neurosurgeon
  (b) Keep the patient on regular follow-up
  (c) Repeat the tests
  (d) Opt for medical management (ketoconazole, mifepristone, and cabergoline)
  (e) Bilateral adrenalectomy

The patient had his tests done several times with appropriate preparation under an ideal setting at a tertiary care hospital. So, there was no doubt regarding his diagnosis of ACTH-dependent Cushing syndrome. Under such circumstances, it would be unnecessary to repeat the tests. Untreated Cushing syndrome has very high mortality/morbidity resulting from CVD events or immunosuppression. So, it would be inappropriate to keep him under follow-up without any definitive measures. Medical management for Cushing syndrome is only adjunctive besides surgery, and none of the drugs have the desired efficacy. As untreated Cushing has a very high mortality (50% in 5 years), our last resort would be bilateral adrenalectomy. However, without localization of the tumor, there is always the possibility of Nelson syndrome if the pituitary tumor produces ACTH. The patient had a recurrent NFA causing significant pressure on surrounding vital structures compelling us to consider TSA.

We consulted a neurosurgeon as the patient had a recurrence of the previous tumor with compression on the optic chiasm that necessitated urgent decompression.

After consultation with the neurosurgeon, our patient immediately underwent repeat pituitary surgery. A few days later, the histopathology and immunohistochemistry reports were available. It was positive for ACTH and FSH immunostaining and reticulin fiber as well as CAM 5.2 (Fig. 22.5). So, our final diagnosis was silent corticotroph adenoma (SCA) that transformed into an active CD following surgery.

One month later, he came for a regular follow-up. He had made a significant subjective and objective improvement. His weight had reduced from 85 to 75 kg, his BP was 100/80 mm Hg without medication, his vision also improved to some extent, and the striae was turning whitish. There was also significant biochemical improvement, and postsurgical MRI showed a small residual tumor (Fig. 22.5d). Finally, he was offered stereotactic radiosurgery (SRS) considering the aggressive course of these types of tumors. Currently, he is on clinical as well as biochemical remission following SRS.

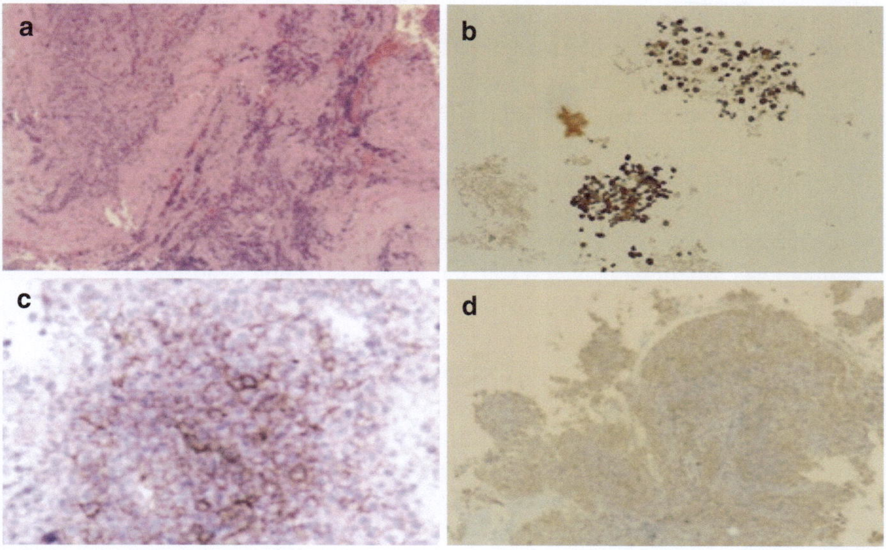

Fig. 22.5 Photomicrographs of the pituitary adenoma showing (**a**) PAS stain; immunopositivity for (**b**) FSH (4+), (**c**) BER EP 4(EP 155), and (**d**) ACTH (2+)

Discussion

Silent corticotroph adenomas (SCAs) are clinically nonfunctioning pituitary adenomas (NFAs) with positive staining for adrenocorticotropic hormone (ACTH) on immunohistochemistry [1, 2]. These adenomas do not exhibit any clinical or biochemical evidence of Cushing's disease. The exact incidence rate of SCAs is unknown, which is reported to vary from 9% to 32.7% [3]. Before surgery, they present in an identical way to NFAs, with clinical features suggesting a local mass effect, including headache, visual field defects, and hypopituitarism [4]. The definitive diagnosis of SCA is determined retrospectively after pathological examination of the resected tumor tissue [5]. Although preoperative features are like those of other NFAs, SCAs tend to have a more aggressive postoperative course [6, 7]. Some studies have demonstrated that SCAs recur earlier, and de novo or multiple recurrences were observed more frequently compared with other NFAs [8, 9]. Under rare circumstances, they can transform into a hormonally active disease presenting with severe Cushing syndrome. They might often produce diagnostic dilemmas with difficult management issues if not thoroughly investigated and subtyped accordingly following speculative surgery.

SCAs are different from CD in the sense that they are clinically and hormonally inactive. Commonly, they are aggressive macroadenomas in contrast to the predominance of microadenoma in the latter. As a result, they are included in the NFA categories that are diagnosed based on immunohistochemical staining. Initially, such tumors present with a mass effect like headache, vomiting, or visual disturbance. However, certain characteristics alert a clinician toward the diagnosis. The patient in our scenario was a young male who presented with compressive symptoms of a large pituitary tumor that had no clinical or biochemical activity. Shortly after undergoing TSA, he developed an aggressive CD with very high ACTH and nonsuppressed HPA axis even on HDDST. Considering the straightforward clinical picture, immunohistochemistry was not advised at the initial operation. So, we were in a diagnostic dilemma between CD and ectopic Cushing syndrome on readmission. Imaging studies failed to localize ectopic ACTH-secreting tumor but showed early recurrence of the invasive pituitary tumor with optic nerve compression demanding immediate TSA. Positive postsurgical immunohistochemical marker of the tumor for ACTH along with significant clinical and biochemical improvement led to the ultimate diagnosis.

Historically, SCA is considered an aggressive tumor characterized by a biologic behavior between that of (ACTH-negative) NFA and adenomas causing CD [10]; the recurrence of or progression of residual tumor in patients with NFA after microsurgical resection varies, depending upon the extent of resection. A recurrence rate of 10–25% was reported in patients in whom a gross total resection was achieved, whereas it was as high as 50% in patients who underwent a subtotal resection [11].

Answer

Q1. (b)
Q2. (c)
Q3. (a)

References

1. Lopez JA, Kleinschmidt-Demasters BK, Sze CI, Woodmansee WW, Lillehei KO. Silent corticotroph adenomas: further clinical and pathological observations. Hum Pathol. 2004;35:1137–47.
2. Cho HY, Cho SW, Kim SW, Shin CS, Park KS, Kim SY. Silent corticotroph adenomas have unique recurrence characteristics compared with other nonfunctioning pituitary adenomas. Clin Endocrinol. 2010;72:648–53.
3. Young WF Jr. Clinically nonfunctioning pituitary adenomas. In: Thapar K, Kovacs K, Scheithauer BW, Lloyd RV, editors. Diagnosis and management of pituitary tumors. Totowa, NJ: Humana Press; 2001. p. 343–51.
4. Scheithauer BW, Jaap AJ, Horvath E, Kovacs K, Lloyd RV, Meyer FB, Laws ER Jr, Young WF Jr. Clinically silent corticotroph tumors of the pituitary gland. Neurosurgery. 2000;47:723–9; discussion 729.
5. Ben-Shlomo A, Cooper O. Silent corticotroph adenomas. Pituitary. 2018;21:183–93.

6. Ioachimescu AG, Eiland L, Chhabra VS, Mastrogianakis GM, Schniederjan MJ, Brat D, Pileggi AV, Oyesiku NM. Silent corticotrophadenomas: Emory University cohort and comparison with ACTH-negative nonfunctioning pituitary adenomas. Neurosurgery. 2012;71:296–303.
7. Jahangiri A, Wagner JR, Pekmezci M, Hiniker A, Chang EF, Kunwar S, Blevins L, Aghi MK. A comprehensive long-term retrospective analysis of silent corticotrophic adenomas vs. hormone-negative adenomas. Neurosurgery. 2013;73:8–17.
8. Cooper O, Ben-Shlomo A, Bonert V, Bannykh S, Mirocha J, Melmed S. Silent corticogonadotrophadenomas: clinical and cellular characteristics and long-term outcomes. Hormon Cancer. 2010;1:80–92.
9. Baldeweg SE, Pollock JR, Powell M, Ahlquist J. A spectrum of behaviour in silent corticotroph pituitary adenomas. Br J Neurosurg. 2005;19:38–42.
10. Webb KM, Laurent JJ, Okonkwo DO, Lopes MB, Vance ML, Laws ER Jr. Clinical characteristics of silent corticotrophic adenomas and creation of an internet-accessible database to facilitate their multiinstitutional study. Neurosurgery. 2003;53:1076–84; discussion 1084.
11. Turner HE, Stratton IM, Byrne JV, Adams CB, Wass JA. Audit of selected patients with nonfunctioning pituitary adenomas treated without irradiation—a follow-up study. Clin Endocrinol. 1999;51:281–4.

A 48-Year-Old-Male with Weight Loss, Palpitations, and Recurrent Hypoglycemia

23

Introduction

The story of a middle-aged male with Graves' toxicosis who started to have recurrent hypoglycemic episodes following the introduction of antithyroid drugs is beautifully described in this case scenario. There is an elaborate discussion on the rational approach to spontaneous hypoglycemia, and in our patient, thorough investigations including a formal supervised fast test established endogenous hyperinsulinism as the probable etiology. A negative pancreatic scan coupled with positive anti-insulin antibody in plasma clinched the ultimate diagnosis of Hirata's syndrome; the self-limiting nature of the disease further supported our conclusion.

Case Scenario

A 48-year-old male presented with weight loss, palpitation, tremor, heat intolerance, and excessive sweating over the last 4 months. He has had a neck swelling for several years. Based on this clinical feature and supported by biochemical investigations, he was diagnosed with Graves toxicosis and started on oral carbimazole 60 mg daily. His symptoms improved immediately on this medication. However, after 2 weeks of treatment, he started to experience frequent episodes of dizziness, headache, and polydipsia, especially on waking up in the early morning. One month after the initiation of treatment, he experienced an episode of severe convulsion with tongue bite, incontinence, and LOC. His blood sugar was found to be 2.2 mmol/L at that time. He rushed to a nearby hospital and resuscitated with I/V glucose. He continued to have recurrent hypoglycemic episodes after discharge from the hospital and finally was compelled to consult us.

He was a well-cooperative gentleman having an anxious look. His BP was 130/80 mm Hg, pulse rate—80/min, and regular. He had bilateral proptosis, lid lag, and lid retraction. There was diffuse thyromegaly without a bruit, and clinically he

S. Jahan, *Endocrinology Case Studies*,
https://doi.org/10.1007/978-981-95-2770-0_23

Table 23.1 SMBG log of the patient showing recurrent hypoglycemia

Day after admission	FBS	2 h ABF	BL	2 h AL	BD	AD
1st	**2.6**	9.9	5.9	6.6	**3.3**	5.0
2nd	**2.4**	10.7	6.8	7.1	**3.7**	5.2
3rd	**3.1**	11	5.5	5.9	**3.6**	6.1
4th	4.3	**3.2**	4.6	6.3	4.4	5.2
5th	**3.6**	7.2	4.1	5.8	5.8	4.6
6th	**3.7**	7.2	7.3	7.8	4.2	5.1
7th	4.9	**3.4**	8.8	9.0	4.6	**3.5**
8th	**3.1**	4.1	6.5	5.9	4.6	8.7

was mildly toxic. He was advised to take frequent small meals based on complex carbohydrates, and following this dietary modification, his hypoglycemic symptoms improved significantly. He was also asked to keep regular self-monitoring of blood glucose (SMBG) log that showed frequent fasting as well as postprandial hypoglycemia, especially at breakfast and dinner (Table 23.1).

Q1. What is the probable type of hypoglycemia that he is experiencing?
  (a) Spurious hypoglycemia
  (b) Factitious hypoglycemia
  (c) Relative hypoglycemia
  (d) Reactive hypoglycemia
  (e) Fasting hypoglycemia

Spurious hypoglycemia results from some technical error in the assay like the absence of fluoride or other inhibitors of glycolysis in the test tube or prolonged standing time before assay. As this patient was having repeated hypoglycemia evidenced on the SMBG log fulfilling the Whipple's triad, spurious hypoglycemia was unlikely. He was a nondiabetic individual and was not on any type of hypoglycemic medication. Detailed history failed to reveal surreptitious use of any kind of hypoglycemic agent, making factitious hypoglycemia less likely. People having uncontrolled diabetes sometimes experience hypoglycemic symptoms well above the hypoglycemic threshold as they are accustomed to high blood sugar, which is called relative hypoglycemia. This nondiabetic individual had symptoms well below the arbitrary hypoglycemic cutoff, so relative hypoglycemia was also less likely. What he was having was spontaneous hypoglycemia, either reactive or post-absorptive. Reactive hypoglycemia is milder, characterized by autonomic rather than neuroglycopenic symptoms, and it tends to occur within 2–4 h of meal intake. On the other hand, fasting hypoglycemia occurs at least 5–6 h after fasting; it is severe with pronounced neuroglycopenic symptoms, and it may lead to convulsion or LOC if untreated. Our patient was having both fasting and reactive hypoglycemia.

Q2. What important points should be looked for in his history and examination?
  (a) Drug history
  (b) Alcohol intake

(c) H/O jaundice, hematemesis, ascites
(d) Hyperpigmentation
(e) Hypogonadism

Spontaneous hypoglycemia occurs in a nondiabetic individual. The commonest causes are certain drugs like insulin, sulphonylurea, quinine, pentamidine, and salicylate among others. Alcohol is often an overlooked cause of hypoglycemia leading to inanition, depletion of hepatic glycogen storage, etc. Organ failure including chronic liver disease, chronic kidney disease, and congestive cardiac failure can often lead to hypoglycemia. However, it is more common in fulminant acute hepatitis rather than in cirrhosis of the liver. So, the recent history of jaundice and hematemesis with hepatic encephalopathy is more likely to cause hypoglycemia than the previous history of cirrhosis. Growth hormones, cortisol, epinephrine, and glucagon are important counter-regulatory hormones involved in glucose homeostasis. Therefore, deficiency of these hormones might lead to hypoglycemia, e.g., Sheehan syndrome, Addison disease, etc. Hyperpigmentation is an important feature of primary adrenal insufficiency pointing toward hypocortisolism with high ACTH. GH deficiency commonly leads to hypoglycemia in infants or children presenting with growth failure. In adults, it could present as part of panhypopituitarism that is characterized by secondary hypothyroidism, hypogonadism, etc. Our patient was healthy without any feature of organ failure or endocrinopathies.

After admission, he had routine hematologic and biochemical investigations showing neutrophilic leukocytosis, normal hepatic and renal functions, fasting hypoglycemia on OGTT, and normal serum electrolyte levels. He had severe biochemical thyrotoxicosis with adequate cortisol and growth hormone response on stimulation. Antithyroid receptor antibody was positive in high titers (Table 23.2).

So, after the exclusion of drugs, endocrinopathies, and critical or chronic illness leading to hypoglycemia, we had to think of other etiologies that would lead to such symptoms in a healthy person.

Q3. What is not the likely diagnosis at this point?
(a) Insulinoma
(b) Nonpancreatic islet cell tumor
(c) Non-insulinoma pancreatogenous hypoglycemia (NIPH)
(d) Alimentary hypoglycemia
(e) Autoimmune hypoglycemia

If an otherwise healthy individual is experiencing spontaneous fasting hypoglycemia, then hyperinsulinemic disorder needs to be considered seriously. Insulinoma, though rare, now seemed a probability in this patient as he was having severe fasting as well as postprandial hypoglycemia. Other entities that might lead to endogenous hyperinsulinism like a mesenchymal tumor and NIPH were also possible given the pancreatic scan was negative. Alimentary hypoglycemia occurs commonly following gastrectomy leading to postprandial or reactive hypoglycemia. It is not usually associated with severe fasting hypoglycemia as in this case. Autoimmune

Table 23.2 Biochemical investigation of the patient showing severe thyrotoxicosis and multiple positive autoimmune markers

Parameter	Obtained value	Reference range
CBC with ESR	Hb—15 g/L, TC of WBC—16,000/cu mm N—70%, L—15%, E—02%, ESR—46 mm in first hour	Hb—15 ± 2 gm/dL, TC of WBC—7 ± 3$10^9$/L ESR—0–10 mm in first hour
SGPT	28 iu/L	14–63 IU/L
Serum creatinine	0.89 mg/dL	0.5–1.3 mg/dL
Basal cortisol	554 nmol/L	138–690 nmol/L
TSH	**<0.004 μIU/mL**	0.4–4 μIU/mL
FT4	**46.5 pmol/L**	9.13–23.8 pmol/L
FT3	**22.0 pgm/mL**	1.8–4.7 pgm/mL
TRAb	**23 miu/mL**	1.5 miu/mL
GH after stimulation	12 ng/dL	Adequate >10 ng/dL
OGTT	FBS—3.6 mmol/L, 2 h PG—6.6 mmol/L	FPG ≤ 5.5 mmol/L, 2 h PG— ≤ 7.8 mmol/L
IAA	**26 U/L**	2.4 U/L

hypoglycemia is mediated by anti-insulin or insulin receptor antibody that sometimes occurs in the background of other autoimmune diseases. Our patient already had an autoimmune disease on treatment, e.g., Graves' disease on thiourea drug.

Q4. What should be the next plan of investigation?
 (a) 48/72 h formal fast test
 (b) Mixed meal challenge test
 (c) 100 g OGTT
 (d) CT scan of abdomen
 (e) Insulin tolerance test

When the initial screening tests for critical illness and endocrinopathies are negative, it becomes necessary to induce hypoglycemia either spontaneously or after a formal 48/72 h supervised fast to reach the diagnosis. Usually in >70–80% of cases, the diagnostic hypoglycemic threshold is attained within 24 h and in >90% of cases in 48 h. Our patient had inappropriately high insulin and C-peptide level with the corresponding low blood glucose level of 2.5 mmol/L after a 24 h formal fast test supporting the diagnosis of endogenous hyperinsulinism (Table 23.3). He had negative history and biochemical screen for offending drugs (insulin/SU), making surreptitious hypoglycemia unlikely. Thin-slice HRCT of abdomen was advised, and scrutiny of the scan in the arterial phase failed to reveal any pancreatic tumor, making insulinoma a less likely etiology. Reactive/postprandial hypoglycemia sometimes occurs in individuals with impaired glucose tolerance or yet undetected type 2 DM who are in a persistent hyperinsulinemic environment prone to develop hypoglycemic symptoms in case of missed snacks or delayed meals; this also less frequently happens in post-gastrectomy patients because of rapid intestinal transit and

Table 23.3 Result of supervised fast test showing severe hypoglycemia with inappropriately high insulin and C-peptide

Parameter	Obtained value	Reference range
Venous plasma glucose	2.2 mmol/L	–
Plasma insulin	224 μIU/L	<3 μIU/L
Plasma C-peptide	5.8 ng/dL	<0.6 ng/dL
Plasma pro-insulin	2.06 pmol/L	<5 pmol/L

dysinsulinism. Such cases are diagnosed by mixed meal challenge test or preferably by 2/3 h or extended OGTT that shows impaired glucose tolerance in early samples and hypoglycemia in later samples. Insulin tolerance test is indicated in the evaluation of the sufficiency of the HPA axis in suspected adrenal failure and children with growth failure.

Q5. What should be the next plan of investigation?
- (a) Anti-insulin antibody
- (b) IGF-1
- (c) IGF 2
- (d) Somatostatin receptor scintigraphy
- (e) Anti-insulin receptor antibody

Insulinoma is a rare cause of endogeneous hyperinsulinism that often leads to severe fasting hyperglycemia. Our patients had relatively milder fasting as well as postprandial hypoglycemia. Though HRCT of abdomen was negative, it is often not possible to exclude an insulinoma by this as most are <2 mm in size. So, other imaging modalities like somatostatin receptor scintigraphy and endoscopic USG might be required which could not be done because of the lack of local availability. The mesenchymal tumor may at times produce spontaneous hypoglycemia by the elaboration of big IGF-2 molecules where serum insulin is appropriately suppressed. It is diagnosed by an elevated IGF2:IGF1 ratio of >10:1. Our patient had markedly high plasma insulin at the time of hypoglycemia, so such a tumor was less likely to be the cause. The pancreatic scan is unremarkable in patients with NIPH who frequently have a preceding history of gastrectomy/dumping syndrome. Finally, the measurement of anti-insulin antibody (IAA) revealed the underlying etiology as Hirata's syndrome. It was found to be significantly elevated, which presumably led to autoimmune hypoglycemia in this patient. The sulphydryl group in carbimazole is thought to break the bond between the A and B chains of endogenous insulin exposing the fragments to helper T-cells which culminate in the formation of an anti-insulin antibody (AIA). This AIA binds with insulin with high affinity, followed by slow dissociation at a time when plasma glucose is falling, causing symptomatic hypoglycemia.

We advised the patient to have a frequent small meal based on complex CHO enriched in protein and fat and supplemented with uncooked cornstarch. He was prescribed tab. Diazoxide which he was unable to tolerate and so discontinued by himself. Interestingly a few days later, his symptoms improved spontaneously, and he stopped having any further hypoglycemic episodes. This is also supportive evidence in favor of our diagnosis because hypoglycemia is often self-limiting in Hirata syndrome with autoresolution in most cases. At present, he is on regular follow-up and doing well.

Discussion

Insulin autoimmune syndrome (IAS) is a rare condition, characterized by spontaneous episodes of hyperinsulinemic hypoglycemia due to the presence of high serum concentrations of insulin autoantibodies (IAA). IAS is also named Hirata's disease after the original description made in 1970 by Yukimasa Hirata and colleagues [1]. IAS is one of the two types of autoimmune hypoglycemia, the other type B being insulin-resistant, which is due to the antibodies against the insulin receptor [2, 3]. The immunogenetic determinants for IAS were identified in the genes of the class II Human Leukocyte Antigen (HLA) system: the disease resulted strongly associated with HLA-DR4 and specifically with DRB1*0406 and, although less significantly, DRB1*0403 and DRB1*0407 [4]. Many different triggers have been advocated to induce the development of IAS when acting on an underlying predisposing background: among them, in the order of importance, there are medications, viruses, and hematological diseases. Importantly, the medications associated with IAS are mostly sulphydryl and reducing compounds, providing a clue regarding the possible pathogenesis of the disease. The proposed mechanism is that these drugs would bind and cleave the sulphydryl bonds between the insulin chains A and B, thus leading to a conformational change in its molecular structure that makes the endogenous insulin more immunogenic [5]. The mean onset time is estimated to be 4 to 6 weeks following the beginning of the offending medication [6]. High binding capacity and low affinity are the specific features of IAA that are capable of inducing IAS. Due to their high binding capacity, IAA is capable of binding several molecules of insulin, resulting in the formation of large antigen–antibody complexes. On the other hand, the low affinity for insulin is responsible for a significant spontaneous dissociation rate, which inappropriately raises unbound insulin concentrations, thus resulting in hypoglycemic episodes, as reported in the literature [7]. IAS-induced hypoglycemia is usually mild, even though cases presenting severe manifestations (up to seizures and coma) have been reported [8]. The measurement of IAA titer is mandatory for the diagnosis of IAS [9] since it is the only lab test that allows a reliable diagnosis of IAS. It is often a diagnosis of exclusion in a patient with pre-existing autoimmunity or culprit drugs when other tests like pancreatic imaging and insulin or sulphonylurea use have been excluded. The prognosis is favorable; most of the cases undergo spontaneous resolution without further recurrence despite negligible management. Approximately 82% of IAS patients underwent spontaneous remission in a revision of 197 patients diagnosed with IAS from 1970 to 1992 [6]. Treatment includes dietary modification, acarbose, diazoxide, steroid, and immunosuppressive agents like azathioprine, rituximab, etc. In intractable cases, plasmapheresis has been applied with variable success.

Answer

Q1. (d), (e)
Q2. (a), (b), (c), (d), (e)
Q3. (d)
Q4. (a), (d)
Q5. (a)

References

1. Hirata Y, Ishizu H, Ouchi N. Insulin autoimmunity in a case of spontaneous hypoglycemia. J Jpn Diabetes Soc. 1970;13:312–20.
2. De Pirro R, Roth RA, Rossetti L, Goldfine ID. Characterization of the serum from a patient with insulin resistance and hypoglycemia. Evidence for multiple populations of insulin receptor antibodies with different receptor binding and insulin-mimicking activities. Diabetes. 1984;33(3):301–4.
3. Lupsa BC, Chong AY, Cochran EK, Soos MA, Semple RK, Gorden P. Autoimmune forms of hypoglycemia. Medicine (Baltimore). 2009;88(3):141–53.
4. Uchigata Y, Kuwata S, Tokunaga K, et al. Strong association of insulin autoimmune syndrome with HLA-DR4. Lancet (London, England). 1992;339(8790):393–4.
5. Hattori N, Duhita MR, Mukai A, Matsueda M, Shimatsu A. Development of insulin antibodies and changes in titers over a long-term period in patients with type 2 diabetes. Clin Chim Acta. 2014;433:135–8.
6. Uchigata Y, Eguchi Y, Takayama-Hasumi S, Omori Y. Insulin autoimmune syndrome (Hirata disease): clinical features and epidemiology in Japan. Diabetes Res Clin Pract. 1994;22(2–3)
7. Folling I, Norman N. Hyperglycemia, hypoglycemic attacks, and production of anti-insulin antibodies without previous known immunization. Immunological and functional studies in a patient. Diabetes. 1972;21(7):814–26.
8. Wang YL, Yao PW, Zhang XT, Luo ZZ, Wu PQ, Xiao F. Insulin autoimmune syndrome: 73 cases of clinical analysis. Chin Med J. 2015;128(17):2408–9.
9. Ismail AA. Testing for insulin antibodies is mandatory in the differential diagnosis of hypoglycemia in nondiabetic subjects. Clin Endocrinol. 2012;76(4):603–4.

A 7-Year-Old Boy with Precocious Puberty and Lower Limb Weakness

24

Introduction

Case scenario: A 7-year-old boy presented with the complaints of voice change and enlargement of phallus over the last 1 year. His parents noticed that his voice had become huskier along with the development of some behavioral change during the previous year. They were alarmed at his rapid phallic enlargement and the premature appearance of axillary and pubic hair being noticeable mostly over the last 6 months. Three months earlier, he suddenly developed bilateral lower limb weakness, followed by urinary incontinence.

He had an uneventful antenatal period without any significant history of birth trauma or any maternal drug ingestion. He was born to a parent of nonconsanguineous marriage. All his family members were healthy without any similar type of illness. He did not report any headache or seizures or exposure to any exogenous sex steroids (medicinal or cosmetic sources).

On examination, he seemed to be somewhat agitated having a masculine body habitus. His height was 135 cm (50th percentile) and weight was 29 kg (75th percentile). Sexual maturation rating by Tanner staging revealed testicular volume bilaterally of 8 mL and abundant pubic hair (P4) with stretched penile length of 9 cm (Fig. 24.1). Neurologic examination of lower limb showed the presence of pes cavus, normal bulk and tone of muscles on both sides, power—4/5 proximally (B/L) and 2/5 on dorsi and plantar flexion, absent ankle jerk, and bilateral plantar extensor with limping gait. There was a symmetrical sensory loss in a saddle distribution. All cranial nerves including fundoscopic examination were unremarkable.

S. Jahan, *Endocrinology Case Studies*,
https://doi.org/10.1007/978-981-95-2770-0_24

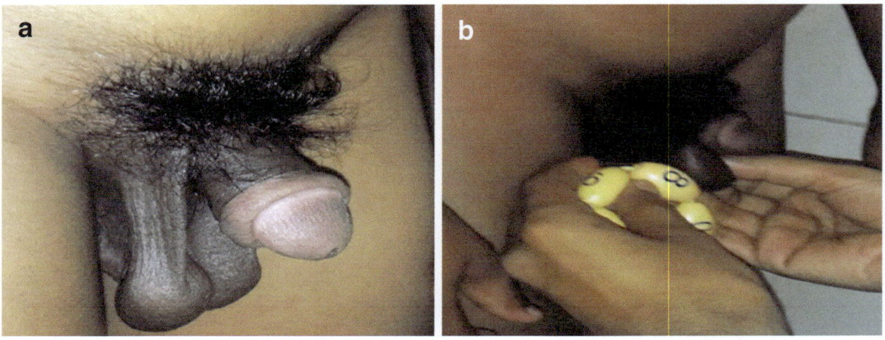

Fig 24.1 (**a** and **b**) Adult appearance of testes and phallus of the boy

Q1. What is the probable diagnosis depending on clinical features?
  (a) Nonprogressive complete precocious puberty (CPP)
  (b) Progressive complete precocious puberty
  (c) Nonprogressive peripheral precocious puberty (GISP)
  (d) Contrasexual precocious puberty
  (e) Progressive peripheral precocious puberty

The appearance of pubertal signs in a boy before 9 years of age is considered precocious puberty, and the underlying etiology is organic in about 75% of the cases. If sexual precocity results from premature activation of the hypothalamic GnRH pulse generator/pituitary gonadotropin–gonadal axis, the condition is GnRH-dependent and is termed central or complete precocious puberty (CPP) or true isosexual precocity (ISP). If extrapituitary secretion of gonadotropins or secretion of gonadal steroids independent of pulsatile GnRH stimulation leads to early virilization in boys or feminization in girls, the condition is termed incomplete ISP, pseudoprecocious puberty, peripheral precocity, or GISP. When the production of increased estrogen levels in males leads to inappropriate feminization and the production of excessive androgen in a female leads to inappropriate virilization, then these conditions are termed contrasexual precocity or heterosexual precocity. The boy was undoubtedly virilized prematurely with a testicular volume of 8 mL (far beyond the pubertal cutoff >4 mL) and the other pubertal signs appearing after gonadarche. This is in favor of GnRH-dependent/complete precocious puberty. History was suggestive of progressive CPP as there was a rapid progression of pubertal stages, accelerated growth velocity, and far advanced bone age with sexual maturation well into the adult range. So, at this stage, the diagnosis was progressive complete precocious puberty. Considering all the facts together, it was likely to be organic in origin rather than idiopathic.

His investigations revealed an advanced bone age (between 12 and 16 years), serum testosterone level well into the adult male range, and pubertal response on GnRH stimulation test supporting CPP. β-HCG, alpha-fetoprotein, and other tumor markers were normal (Table 24.1). Ultrasonography of the testis showed bilateral symmetric testicular enlargement. Pituitary and CNS MRI was unremarkable.

Table 24.1 Biochemical investigation of the patient showing markedly elevated serum testosterone and pubertal response in the GnRH stimulation test

Biochemical parameter	Patient's value	Reference range
Serum total testosterone	**23.73 nmol/L**	<0.45 nmol/L
GnRH-stimulated LH	**12.6 mIU/mL**	0.00–0.10 mIU/mL
GnRH-stimulated FSH	8.38 mIU/mL	<3.00 mIU/mL
Plasma β-HCG	1.9 mIU/mL	0.00–2.5 mIU/mL
Plasma α-fetoprotein	12 ng/mL	10–20 ng/mL

Q2. What could be the probable etiology of CPP in this boy?
  (a) β-HCG-secreting tumor
  (b) Leydig cell tumor
  (c) Congenital adrenal hyperplasia
  (d) Familial testotoxicosis
  (e) CNS tumor (optic glioma, ependymoma), etc

Boys with β-HCG-secreting neoplasms have slightly enlarged testes (although not to a size consistent with the size of the phallus and other male secondary sex characteristics because the seminiferous tubules are not affected), and it may be difficult to differentiate these patients from boys in the early stages of CPP based on physical examination alone. However, plasma β-HCG levels are elevated without an increase in the concentration of FSH or LH. The premature appearance of pubic hair, phallic enlargement, and other signs of virilization in a male without enlargement of the testes or the liver suggests the diagnosis of congenital virilizing CAH and virilizing adrenal tumor. In CAH, plasma 17-hydroxyprogesterone, 11-deoxycortisol, and corticosterone will be elevated depending on the site of the enzyme block in the steroid synthesis pathway. Such a child is likely to be born of consanguineous marriage, having a history of salt-wasting crisis and hyperpigmentation on examination. There might be a palpable abdominal lump with signs of hormonal excess (glucocorticoid and mineralocorticoid) in virilizing adrenal tumor, and DHEAS concentrations are likely to be very high. Leydig cell tumor is commonly a malignant neoplasm leading to irregular, asymmetric enlargement of testes with very high testosterone and suppressed LH having no response to GnRH stimulation. Familial testotoxicosis is an autosomal dominant disorder characterized by the testes that are symmetric and > 2.5 cm but smaller than that expected for the stage of pubertal development. These patients have pubertal values of testosterone and suppressed LH with no response to GnRH stimulation. CPP due to CNS tumor was the most likely diagnosis in this boy as there was significant, symmetric testicular enlargement occurring earlier and being consistent than other signs of sexual maturation. Negative biochemical screen (β-hCG and adrenal steroid) and imaging studies (ultrasound of testes and abdomen) were also supportive of this diagnosis. However, MRI failed to localize any CNS lesion, and the patient had no signs of raised ICP or any localizing features. So, we were in a state of dilemma at this point.

Q3. What other investigations are indicated in this patient?
(a) MRI/CT scans of chest and abdomen
(b) Testicular biopsy
(c) MRI of the spine
(d) CSF study
(e) None of the above

Sometimes, imaging of the chest or abdomen in a child may lead to the diagnosis of a germ cell tumor or ectopic β-HCG-secreting tumor. However, these tumors lead to GnRH-independent precocious puberty by activating the LH receptor in the Leydig cell. Rarely, when located near the hypothalamus, they may lead to CPP by impinging on the GnRH pulse generator. In this case, such imaging did not yield any diagnosis. A testicular biopsy would be helpful in Leydig cell tumor, eugonadal germ cell tumor, or testicular adrenal rest tissue (TART). All of these were also less likely in this boy as they are associated with GISP rather than CPP. Sometimes, CSF is analyzed in cases of sexual precocity to determine the level of tumor markers like β-HCG or α-fetoprotein suggestive of β-HCG-secreting or germ cell tumor. However, these were unlikely to be helpful in this boy. Though there was no apparent relationship between the neurological signs (urinary incontinence and spastic paraparesis) and precocious puberty, nonetheless the MRI of spine was indicated considering the clinical signs.

MRI of spine revealed intramedullary lesion at the lower part of the spinal cord extending up to conus medullaris (Fig. 24.2). The patient underwent total resection

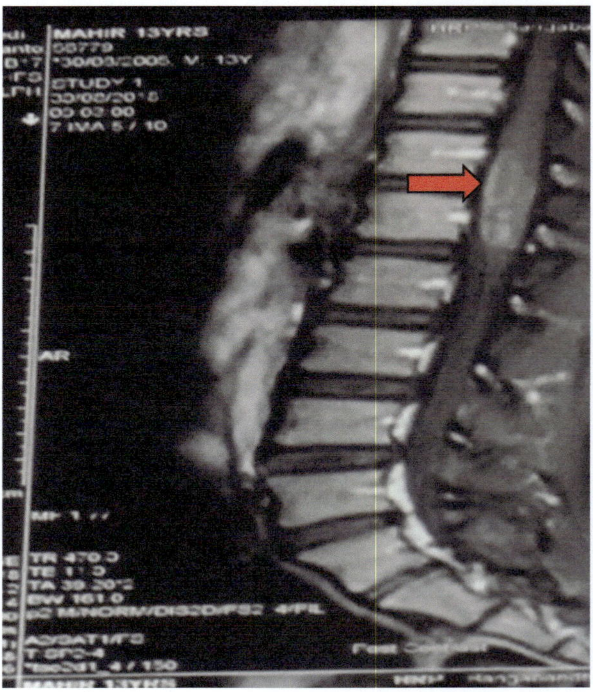

Fig. 24.2 MRI of spine showing intraspinal lesion in conus medullaris

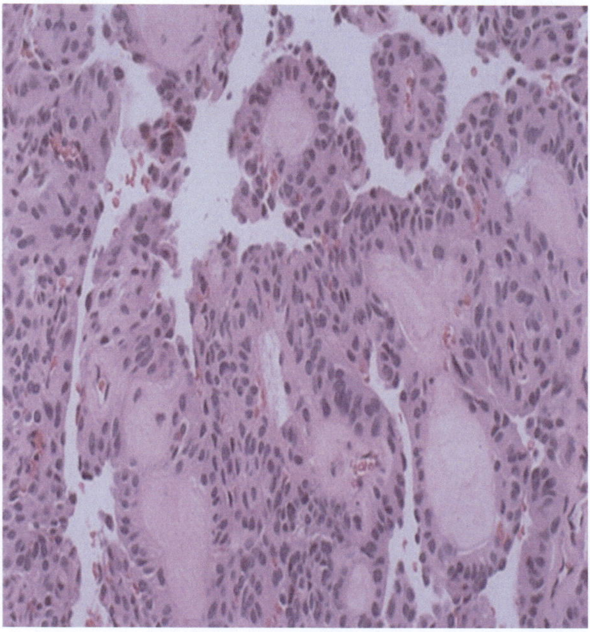

Fig. 24.3 Histopathology showing benign ependymoma

of the tumor, and histological (Fig. 24.3) examination revealed a benign ependymoma. Immunostaining for gonadotropins was found to be positive. At 6 months of follow-up, there was significant regression of the features of precocious puberty. He is currently on physiotherapy and doing well.

Discussion

Central precocious puberty (CPP) which is caused by premature activation of the hypothalamic–pituitary–gonadal (HPG) axis is more common in girls than in boys [1]. More than 50% of boys with CPP have an identifiable etiology, whereas in girls, it is only 20% [1]. Therefore, CNS disease needs to be ruled out before reaching a diagnosis of idiopathic precocious puberty in a boy. Tumors (optic glioma, astrocytoma, ependymoma, and medulloblastoma), infections, congenital defects, cerebral ischemia, and radiation or injury to the CNS can all lead to secondary CPP [2]. Though CNS tumors are relatively common childhood neoplasms, tumors presenting with precocious puberty are relatively uncommon [3]. Several CNS tumors contributing to precocious puberty have been described. Commonly, these tumors are in the sellar and/or suprasellar region of the brain, thereby directly disrupting the normal prepubertal inhibition of the HPG axis. Sometimes, tumors distant from the sella may indirectly cause GnRH stimulation through pressure on the hypothalamic–pituitary region from concurrent hydrocephalus [4].

CNS ependymoma is one such rare cause of CPP. It accounts for ~10% of CNS tumors in children. The incidence is 2.2 cases per million and appears to be increasing over the past 30 years [5]. It is the third most common central nervous system (CNS) tumor in childhood, most often located intracranially, with a predilection for the posterior fossa [6, 7]. ***Spinal ependymomas*** are exceedingly rare in children and adolescents in contrast to adults. Presentation occurs earlier than in adults with compressive symptoms, whereas adults most often present with radiculopathy. The patient scenario presented with features of CPP and compressive neurological features. He was treated with total surgical resection leading to complete recovery. It is postulated that intracranial ependymomas compress the inhibitory pathway to the GnRH pulse generator by producing hydrocephalus. But being in the spinal cord, this mechanism is not appropriate here. Rather, production of gonadotropins might be the answer. In that case, this will probably be the first case report of ependymoma producing CPP by secreting gonadotropins. Resolution of CPP after removal of the tumor provided indirect evidence of its association with CPP.

Complete resection alone in spinal cord ependymoma can achieve excellent local control and survival. Patients should receive complete resection if technically possible. Postoperative radiotherapy is not recommended for complete resection. For incomplete resection, postoperative local radiotherapy is recommended, and it can also achieve excellent local control and survival [6].

Answer

Q1. (b)
Q2. (e)
Q3. (c)

References

1. Styne DM. Puberty: ontogeny, neuroendocrinology, physiology, and disorders. In: Melmed S, et al., editors. Williams textbook of endocrinology. 12th ed. Philadelphia: Elsevier Saunders; 2011. p. 1143). John F. Kennedy Blvd.
2. Wilne S, Collier J, Kennedy C, Koller K, Grundy R, Walker D. Presentation of childhood CNS tumours: a systematic review and meta-analysis. Lancet Oncol. 2007;8(8):685–95.
3. Josan VA, Timms CD, Rickert C, Wallace D. Cerebellar astrocytoma presenting with precocious puberty in a girl. Case report. J Neurosurg. 2007;107(1):66–8.
4. Nejat F, El Khashab M, Rutka JT. Initial management of childhood brain tumors: neurosurgical considerations. J Child Neurol. 2008;23(10):1136–48.
5. Helseth E, Due-Tønnessen B, Lote K, Skullerud K, Storm-Mathisen I, Wesenberg F, et al. Ependymoma in children and young adults (0–19 years): report of 25 consecutive cases. Childs Nerv Syst. 2001;17:24–30.
6. McGuire CS, Sainani KL, Fisher PG. Both location and age predict survival in ependymoma: a SEER study. Pediatr Blood Cancer. 2009;52:65–9. https://doi.org/10.1002/pbc.21806.
7. Peris-Bonet R, Martinez-Garcia C, Lacour B, et al. Childhood central nervous system tumours—incidence and survival in Europe (1978–1997): report from automated childhood cancer information system project. Eur J Cancer. 2006;42:2064–80.

A 30-Year-Old Female with Fever, Cough, and an Incidental Adrenal Mass

25

Introduction

This is the sad story of a young lady who was evaluated for an adrenal incidentaloma found on USG of abdomen. Initially, she was found negative for pulmonary TB when investigated based on fever, weight loss, and dry cough. Following admission into our unit, she was reported to have fluctuating HTN with target organ damage and admitted of having a typical Pheo spell. Interestingly, she also had a moonface, hyperpigmentation, prominent proximal myopathy, and hypokalemia. Biochemical and radiologic investigations were in favor of an ACTH-producing pheochromocytoma, where the immunosuppressed state favored opportunistic infection like TB that proved positive on retesting. She was discharged on anti-TB and antihypertensive medication with a future plan of surgery; unfortunately, she died soon after following a Pheo spell resulting in pulmonary edema.

Case Scenario

A 30-year-old garment factory worker presented with an evening rise in temperature associated with dry cough and significant weakness for 6 months. At times, she noticed expectoration of sputum streaked with blood. She also complained about the darkening of her complexion over the last 1 month. On query, she admitted to having recurrent paroxysms of headache with palpitations and sweating for the past 3 months. Each attack lasted for a few minutes to several hours. It was triggered by stress or eating and even tended to occur spontaneously multiple times a day, which was associated with tremors, anxiety, and epigastric pain. Her previous investigations showed neutrophilic leukocytosis with an elevated ESR and a right-sided consolidation with a cavitary lesion on CXR (Fig. 25.1). Sputum for Z-N staining for AFB was found to be negative (Fig. 25.2). USG of abdomen incidentally detected a

S. Jahan, *Endocrinology Case Studies*,
https://doi.org/10.1007/978-981-95-2770-0_25

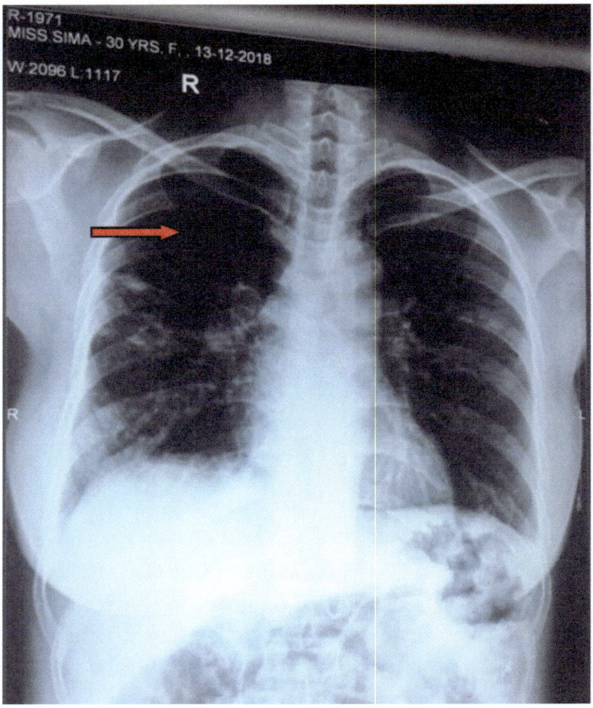

Fig. 25.1 CXR P/A view showing cavitary lesion

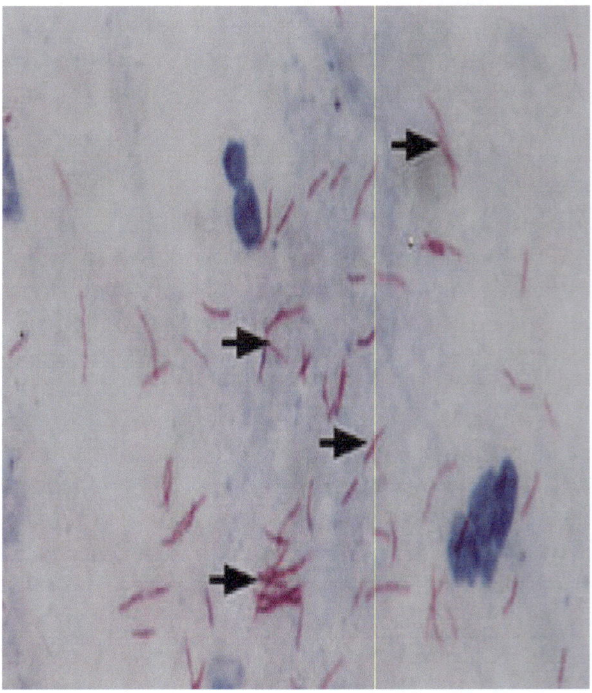

Fig. 25.2 Sputum smear showing AFB

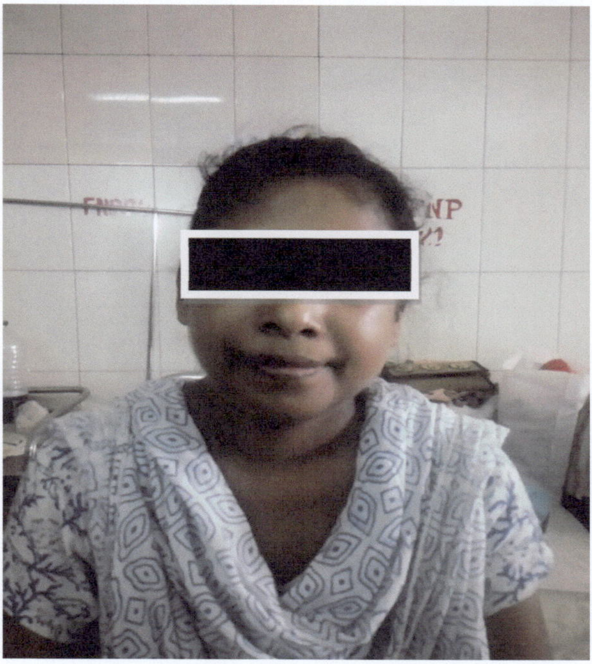

Fig 25.3 Typical moon face of the patient

left-sided adrenal mass of (5.7 × 4.6) cm which compelled her admission into our unit for further evaluation.

On examination, she had a typical moon face, generalized hyperpigmentation more so on palmar creases and knuckles (Figs. 25.3 and 25.4). She also had significant proximal myopathy and grade IV hypertensive retinopathy. Ambulatory BP monitoring revealed marked fluctuating HTN (Table 25.1).

Q1. What are the possible diagnoses based on these information?
  (a) Pheochromocytoma
  (b) Disseminated tuberculosis
  (c) Disseminated histoplasmosis
  (d) Adrenal metastasis
  (e) Adrenocortical carcinoma (ACC)

Considering the fact that she had a typical Pheo spell (a classic triad of headache, palpitations, and sweating), young-onset HTN, severe and fluctuating in nature along with end-organ damage (hypertensive retinopathy) and an adrenal incidentaloma, made Pheo a high probability. Low-grade fever and her chest symptoms, consolidation, and cavitary pulmonary lesion could be explained by TB or histoplasmosis with dissemination to the adrenals. A cavitary primary tumor in the lungs, with secondary in adrenals, would be a possibility though her young age was against this diagnosis. Likewise, she might have primary adrenal carcinoma with pulmonary metastasis. However, 60% of ACC is hormonally active and presents with the

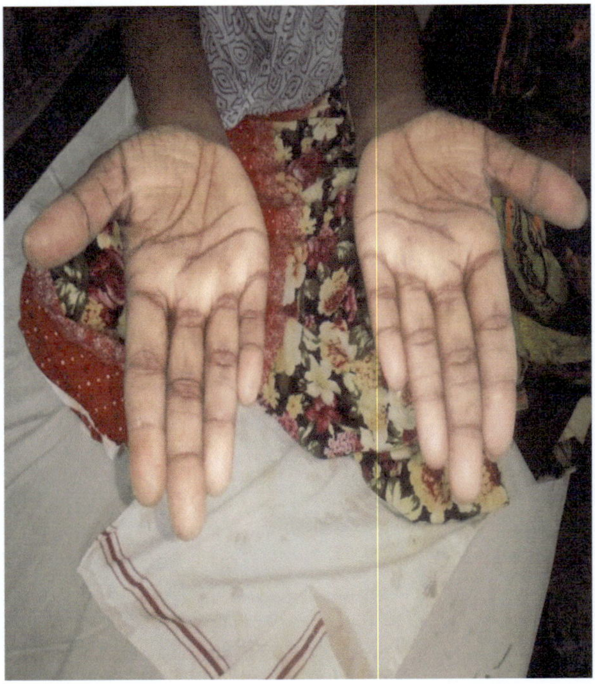

Fig. 25.4 Hyperpigmentation of palmar creases

Table 25.1 Chart of ambulatory BP monitoring (in mm of Hg) showing fluctuating HTN

Date	8 am	12 pm	4 pm	8 pm
6.7.19	110/70	**220/140**	140/100	150/100
7.7.19	170/100	190/130	180/120	190/140
8.7.19	210/130	110/75	215/140	140/95
9.7.19	170/130	110/80	160/100	**85/60**

feature of Cushing syndrome and/or virilization due to glucocorticoid or androgen hypersecretion. However, none of the differentials other than pheochromocytoma were probably enough to explain the severe, fluctuating HTN and the recurrent panic spells.

After admission, relevant investigations were carried out, and results showed marked polycythemia with neutrophilic leukocytosis, moderate hypokalemia, hypovitaminosis D leading to secondary hyperparathyroidism, and incidentally detected uncontrolled DM. She also had possible hypertensive and diabetic nephropathy as evidenced by glycosuria and proteinuria on routine urine examination (Table 25.2).

Q2. What might be the possible reason behind her significant hypokalemia?
   (a) Conn syndrome
   (b) Renal tubular acidosis (RTA) secondary to hyperparathyroidism
   (c) Diabetic kidney disease (DKD)

Table 25.2 Routine biochemical investigation of the patient showing hypokalemia and hyperglycemia

Test	Result	Reference range
CBC and ESR	Hb%—14.3 g/dL, ESR—35 mm in first h, RBC—**5.04 × 10 ^{12}/L** WBC—**21 × 10 ^{9}/L** TC of WBC: N—84%, L—11% Platelet—470 × 10^9/L	Hb—13.5 ± 1.3 g/dL, ESR—0–10 mm in first h RBC—4.3 ± 0.5 × 10^{12}/L TC of WBC—7 ± 3 × 10^9/L Platelet—150–400 × 10^3/L
Serum creatinine	1.1 mg/dL	0.5–1.3 mg/dL
Serum electrolyte	Na—137, K—**2.9 mmol/L**	Na—135–145, K—3.5–5.1, Cl—95–107, TCO_2—22–30 mmol/L
Serum albumin	46 gm/L	34–54 g/L
Serum calcium	9.9 mg/dL	8.5–10 mg/dL
Serum PO4	4.1 mg/dL	4–7 mg/dL
Plasma iPTH	**130 pg/mL**	11–67 pg/mL
Serum 25(OH) Vit-D	17 ng/dL	20–40 ng/dL
Serum SGPT	29 IU/L	14–63 IU/L
FBS	**21.2 mmol/L**	≤5.5 mmol/L
2 h PG	**28.9 mmol/L**	≤7.8 mmol/L
HbA1c	7.7%	≤6.4%

(d) Pheochromocytoma
(e) Cushing syndrome (CS)

Though Conn syndrome can lead to hypokalemic HTN, it is not associated with typical panic spells suggestive of Pheo. At the same time, the adrenal mass was large and heterogeneous, which is not common for a Conn's adenoma. RTA is a well-established cause of recurrent hypokalemia. However, patients with RTA are generally normotensive having systemic acidosis rather than alkalosis as in this case. The secondary hyperparathyroidism resulting from hypovitaminosis D does not lead to RTA which is unlike primary hyperparathyroidism. The duration of her illness was very brief to explain DKD which commonly occurs in long-standing, uncontrolled DM, and the absence of diabetic retinopathy made it an unlikely diagnosis. Pheochromocytoma itself does not lead to any type of electrolyte imbalance. On the other hand, Cushing syndrome, specifically ectopic CS, can lead to hypokalemic HTN with alkalosis. The typical moon face, proximal myopathy, hyperpigmentation, hyperglycemia all could be explained by the presence of ectopic CS.

Endocrine investigations revealed that she had significantly high 24 h urinary metanephrine favoring the diagnosis of pheochromocytoma. She also had a marked elevation of 24 h urinary free cortisol. Her basal plasma cortisol as well as her

Table 25.3 Hormonal investigations of the patient showing high plasma cortisol/ACTH and urinary cotisol and metanephrine

Test	Result	Reference value
FT4	1.31 ng/dL	0.7–1.9 ng/dL
TSH	3.3 mIU/L	0.5–5.1 mIU/L
Plasma ACTH	**562 pg/mL**	0–46 pg/mL
Basal cortisol	**1258 nmol /L**	139–460 nmol/L
24 h UFC	**655 μgm/day**	50–190 μgm/day
24 h urinary metanephrine	**3291 nmol/d**	< 1775 nmol/day
Serum calcitonin	7.10 pgm /mL	11.5 pgm /mL

plasma ACTH were also elevated. Other hormonal parameters were unremarkable (Table 25.3).

Q3. What could be the reason behind her high plasma and urinary cortisol?
- (a) Cushing disease
- (b) Ectopic Cushing syndrome
- (c) Adrenal Cushing syndrome
- (d) ACTH-secreting pheochromocytoma
- (e) Pseudo-Cushing syndrome

Typical clinical phenotype (moon face, proximal myopathy, hypokalemic HTN, and uncontrolled DM) with >3 times elevation of UFC (done on two occasions) was enough to establish the diagnosis of Cushing syndrome. The concurrent presence of the metabolic and catabolic features of Cushing syndrome along with the marked increase of UFC made pseudo-Cushing syndrome an unlikely diagnosis. Cushing's disease is typically a less severe form of hypercortisolism than ectopic CS presenting with hyperandrogenic features (hirsutism and oligomenorrhea) in a young female as in this case though she lacked any hyperandrogenic stigmata. However, the very short history of severe CS and its rapid progression made it less probable. At the same time, hypokalemia and hyperpigmentation are less likely in Cushing's disease, and plasma ACTH is usually mildly elevated/inappropriately normal unlike in our patient having very high ACTH level. All of her features including biochemical pictures could be explained by ectopic CS. Pheochromocytoma is a neuroendocrine tumor elaborating a plethora of peptide hormones including ACTH. So, this lady had pheochromocytoma that was co-expressing ACTH. Adrenal CS leads to less severe CS if not due to ACC which is an unlikely diagnosis here as the adrenal medullary and cortical tumor cannot coexist in the same patient except in rare instances. Again, adrenal CS leads to suppressed rather than elevated ACTH as in her case. Considering all these factors together, she most likely had pheochromocytoma with co-secretion of ACTH which is reported to be present in 5.2% of cases in the literature [1, 2].

Finally, an adrenal CT scan was done that showed a large, well-defined, heterogeneously enhancing left adrenal mass of (5.2 × 4.8 × 5.0) cm having a precontrast

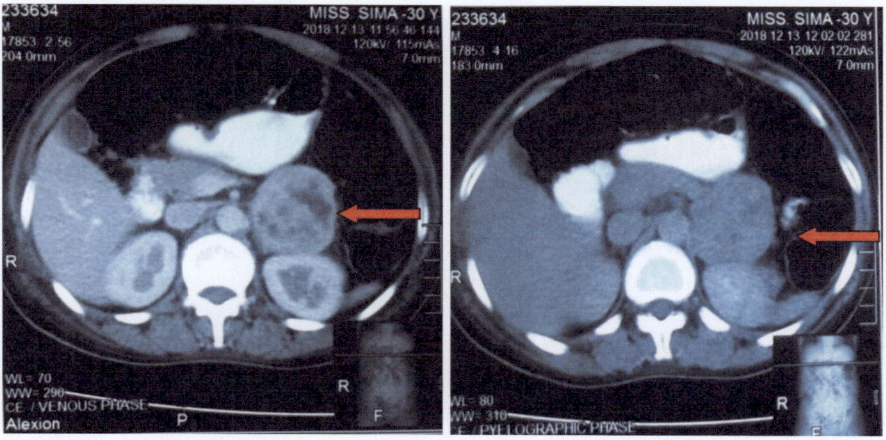

Fig. 25.5 CT adrenals showing a large, heterogeneous left-sided adrenal mass suggestive of pheochromocytoma

density of 56 HU; AWR was 34% at 10 min, and it was strongly suggestive of pheochromocytoma (Fig. 25.5). Her sputum sample was re-examined, and this time, it stained positive for AFB in all three samples. Sustained hypercortisolism can lead to immunosuppression, thereby favoring opportunistic infections like TB.

Q4. What should be the next management plan?
- (a) Resecting the tumor immediately
- (b) Resection of tumor after initiation of appropriate antihypertensive medication
- (c) Planning surgery after becoming AFB-negative on anti-TB medication
- (d) Planning surgery after completion of anti-TB treatment
- (e) Treaingt her medically with regular follow-up

This lady probably had malignant pheochromocytoma having a size of >5 cm that portends a grave prognosis with a mortality rate of ≥43%. Such a case should be operated upon immediately after initiation and titration of antihypertensive medication, e.g., α- and β-blockers sequentially. Neither resection of the tumor before HTN control nor waiting too long for the completion of either the initiation or maintenance phase of anti-TB treatment would be appropriate as both might lead to a catastrophe. Medical treatment with follow-up is not very effective and is only advised in malignant, inoperable cases.

Our patient was put on α- and β-blockers, insulin, and anti-TB medication. She was referred to an endocrine surgeon who planned to operate on her after completion of the initiation phase of antitubercular therapy. Unfortunately, a few days later, at her home, she developed a hypertensive crisis and died of acute pulmonary edema.

Discussion

Cushing syndrome is a rare endocrine disorder characterized by hypercortisolemia. Pituitary sources of ACTH production or Cushing disease comprise about 70% and adrenal tumor another 20% of total cases of CS. Ectopic CS is the rarest entity contributing to only 10% of endogenous CS. Most of the cases of ECS result from bronchial carcinoid. ***Ectopic CS*** caused by ***ACTH-producing pheochromocytomas*** has been reported in only 25 patients so far [1, 2]. Amalgamating the data from six large series of patients with ectopic ACTH, pheochromocytoma was the source of ACTH secretion in 19 out of 363 patients (5.2%) [3]. ACTH-secreting pheochromocytomas form a distinctive subset of patients with CS: the mean age of presentation was 46.8 (range 25–74) years, being more common in females. Most of these patients suffer from severe CS: excessive cortisol secretion resulted in hypokalemia in over 70% and diabetes mellitus in 90% of the cases [4, 5]. In contrast, the prevalence of hypokalemia in pituitary-dependent Cushing's disease is only about 20% [5, 6]. The tests which may help in biochemically confirming ECS are failure of suppression of cortisol levels with 48 h high-dose dexamethasone (8 mg/ day) or poor or absent cortisol and/or ACTH response to intravenous corticotrophin-releasing hormone (CRH). However, there are many exceptions, and the highest diagnostic accuracy is provided by bilateral inferior petrosal sinus sampling, where the ratio of ACTH levels in central: peripheral venous circulation is used to identify a pituitary source for ACTH [4]. However, a suggestive clinical presentation and biochemistry, with clear imaging suggestive of an ectopic source, may obviate the need for this relatively invasive investigation. Venous sampling was utilized in 10 of the 25 (40%) reported patients of ACTH-producing pheochromocytoma. In most patients, invasive venous sampling may be avoided by applying the criteria of Forman et al. [2]. If diagnosed and operated timely following adequate α- and β-blockade, the outcome is optimal; however, if left untreated or in case of treatment delay, the prognosis is fatal.

Answer

Q1. (a)
Q2. (e)
Q3. (b)
Q4. (b)

References

1. Isidori AM, Kaltsas GA, Pozza C, Frajese V, Newell-Price J, Reznek RH, Jenkins PJ, Monson JP, Grossman AB, Besser M. The ectopic adrenocorticotropin syndrome: clinical features, diagnosis, management and long-term follow-up. J Clin Endocrinol Metab. 2006;91(2):371–7.

2. Nijhoff MF, Dekkers OM, Vleming LJ, Smit JWA, Romijn JA, Pereira AM. ACTH-producing pheochromocytoma: clinical considerations and concise review of the literature. Eur J Intern Med. 2009;20:682–5.
3. Etxabe J, Vazquez JA. Morbidity and mortality in Cushing's disease: an epidemiological approach. Clin Endocrinol. 1994;40:479–84.
4. Torpy DJ, Mullen N, Ilias I, Nieman LK. Association of hypertension and hypokalemia with Cushing's syndrome caused by ectopic ACTH secretion. Ann N Y Acad Sci. 2002;970:134–44. [5]
5. Howlett Drury PL, Perry L, Doniach I, Rees LH, Besser GM. Diagnosis and management of ACTH dependent Cushing's syndrome: comparison of the features of ectopic and pituitary ACTH production. Clin Endocrinol. 1986;24:699–713.
6. Pecori Giraldi F, Moro M, Cavagnini F, Study Group on the Hypothalamo–Pituitary–Adrenal Axis of the Italian Society of Endocrinology. Gender-related differences in the presentation and course of Cushing's disease. J Clin Endocrinol Metab. 2003;88(4):1554–8.

An Eight-Year-Old Boy with Behavioral Abnormality and Phallic Enlargement

26

Introduction

This case scenario highlights the unique co-occurrence of hypothalamic hamartoma with neurofibromatosis-1 in an 8-year-old boy leading to central sexual precocity. He presented with well-advanced puberty and behavioral problems along with the stigmata of NF-1 including café-au-lait' macules and skeletal deformities. Biochemical investigations suggested that CPP is commonly due to optic nerve glioma in NF-1. However, he had dual pathology where CPP resulted from associated hypothalamic hamartoma in the absence of optic glioma. He was treated with GnRH agonist resulting in successful regression of the symptoms.

Case Scenario

An eight-year-old boy presented with phallic and testicular enlargement for 5 years and behavioral abnormality for the last 2 years. He was the fifth issue of a nonconsanguineous marriage. His mother had difficult-to-control HTN during pregnancy that was complicated by oligomenorrhea and intrauterine growth retardation (IUGR). He was delivered at term weighing 1.5 kg, followed by an uneventful perinatal period. There was no apparent congenital abnormality like cleft lips, palate, or enlarged phallus at birth. His parents could not recall any episode of neonatal jaundice, hypoglycemia, or any salt-wasting crisis in infancy. However, he had frequent episodes of RTI (pneumonia) requiring hospitalization five to six times till 2 years of age. According to the parents, they first noticed the appearance of sparse pubic hair and progressive enlargement of the penis and testis when the boy was about 3 years of age. At the same time, they noticed that he seemed taller than his peers and appeared to be growing faster than other boys of his age. Around 2 years back, he (when 6 ½ years old) started showing some abnormal behavior. He became aggressive, wanted to establish his own opinion, and showed inappropriate rudeness

S. Jahan, *Endocrinology Case Studies*,
https://doi.org/10.1007/978-981-95-2770-0_26

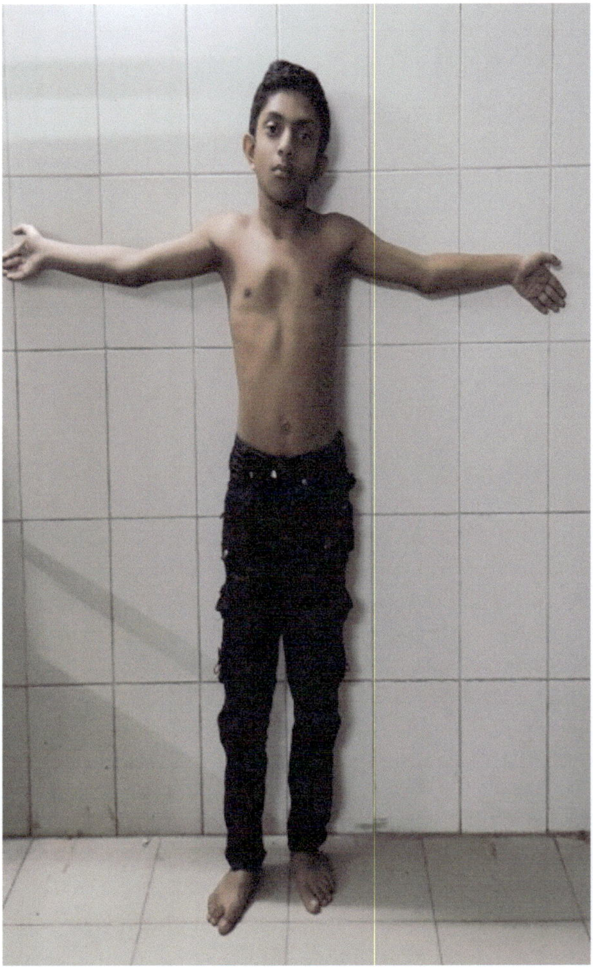

Fig. 26.1 Height acceleration of the boy

with frequent cursing. A few times he was found to be watching explicit content on the smartphone and occasionally demanded that he be married.

These behavioral and physical abnormalities continued to escalate with time. However, during the last year, these changes have slowed down a bit. On query, parents informed that he had some hyperpigmented lesions since birth that gradually increased in size with increasing age.

On examination, he had dysmorphic facies with low-set ears, bilateral short fourth metatarsal, pectus excavatum, and kyphoscoliosis (Figs. 26.1, 26.2, and 26.6). There were multiple smooth-bordered café-au-lait spots on both sides of his body (Figs. 26.4 and 26.5). His height was 134 cm (at 75th centile), weight: 26 kg (at 50th centile), and BMI: 14.3 kg/m^2 (11th centile); his upper segment was 68 cm, while the lower segment was 66 cm; the upper to lower segment ratio was 1.03, and arm span was 132 cm (Figs. 26.1, 26.2, and 26.3). Sexual maturation rating by

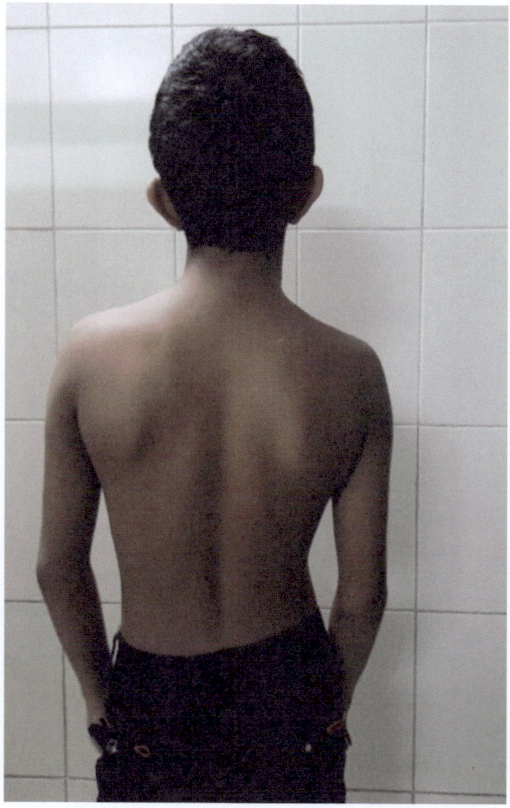

Fig. 26.2 Presence of kyphoscoliosis

Tanner staging revealed well-advanced puberty (testicular volume: right: −12+ cm, left 12+ cm, pubic hair: P4, SPL: 10 cm, scrotal rugosity and pigmentation present) (Fig. 26.7).

Q1. What is the possible diagnosis based on clinical features?
   (a) Idiopathic central precocious puberty (CPP)
   (b) McCune Albright syndrome (MCA)
   (c) Neurofibromatosis-1(NF-1)
   (d) β-HCG-producing tumor
   (e) Hypothalamic lesion (hamartoma, astrocytoma, etc.)

This boy had definite progressive precocious puberty, as evidenced by height acceleration, increased growth velocity, advanced bone age, and inappropriately advanced stage of puberty (Fig. 26.8). In contrast to girls where the majority of CPP is idiopathic, it is only about 25% in the case of boys. He had some nonendocrine manifestations (café-au-lait spot, skeletal deformity) that could not be explained either by idiopathic central precocious puberty or by any hypothalamic tumor.

Fig. 26.3 Evidence of pectus excavatum

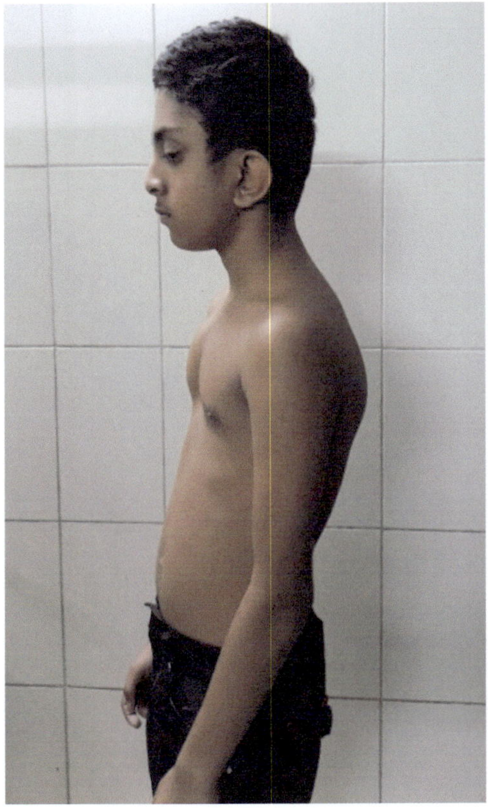

Fig. 26.4 Multiple café-au-lait spots crossing midline

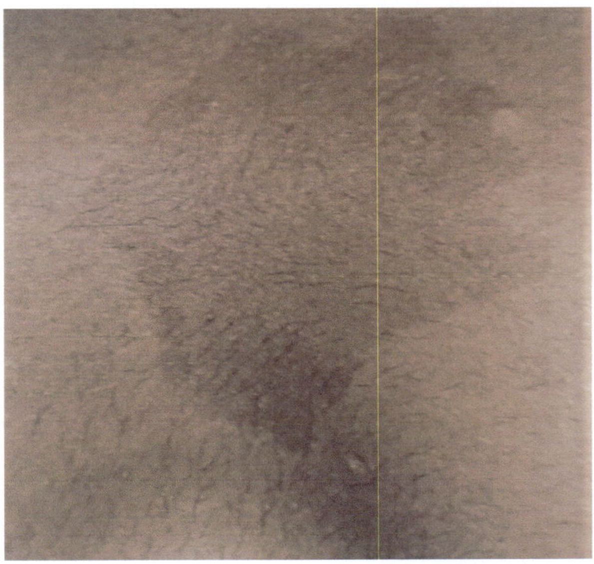

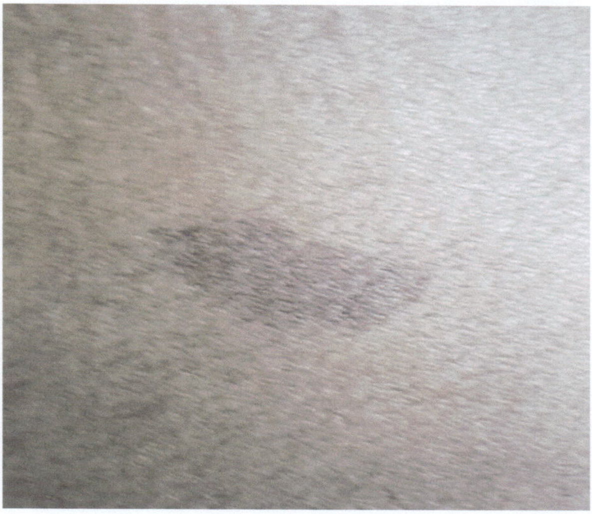

Fig. 26.5 Multiple café-au-lait spots crossing the midline

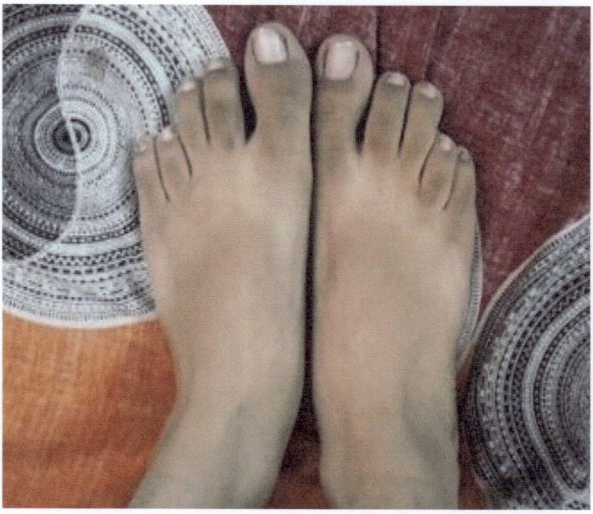

Fig. 26.6 Short 4th and 5th meta-tarsal of the toes

Features related to hypothalamic tumors like a gelastic seizure, headache, impaired vision, hypopituitarism, hyperphagia, or obesity were prominently absent, though they might not be present universally. β-HCG-secreting tumor acts on the LH-CGR receptor on testicular leydig cells leading to peripheral precocity. This boy initially had testicular enlargement, followed by the appearance of pubic hair which usually occurs in reverse sequence in peripheral precocity. Moreover, such a tumor, when eugonadal, causes asymmetric testicular enlargement unlike in this case. In CPP (hypothalamic tumor, idiopathic), the testicular enlargement is more pronounced than phallic enlargement or the stage of pubic hair development, and the opposite is true for peripheral precocity (MCA syndrome, β-HCG-producing tumor). However,

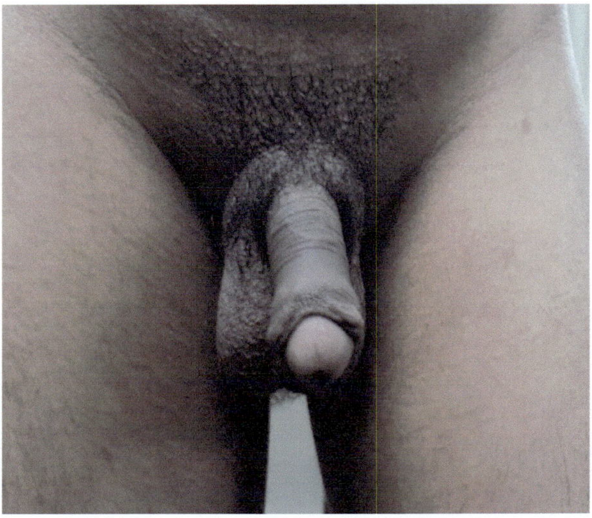

Fig. 26.7 Testicular and phallic enlargement

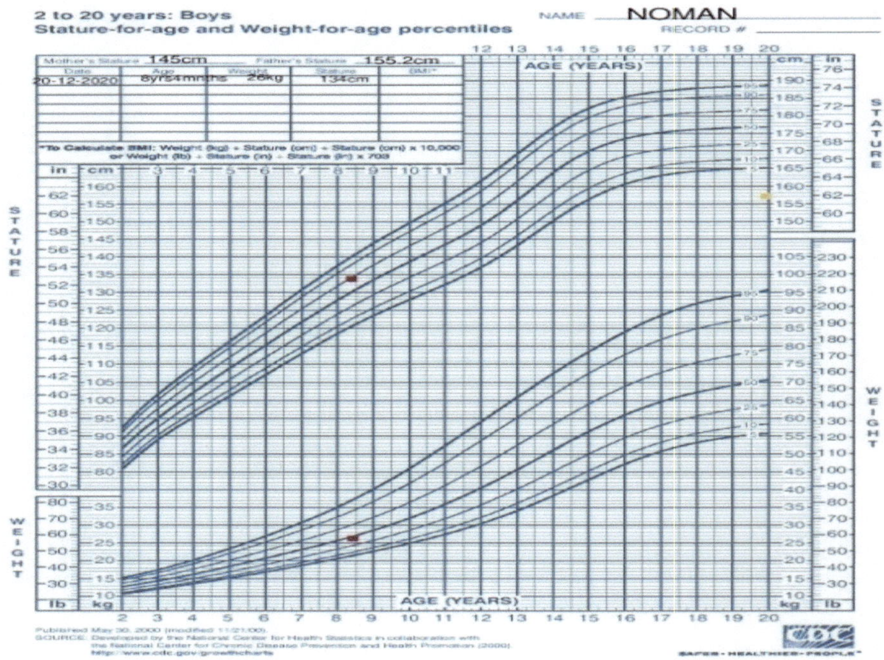

Fig. 26.8 Growth chart of the boy showing height acceleration

Table 26.1 Hormonal investigations of the boy showing central sexual precocity

Date	14-11-2020	Reference range
TSH	3.54uIU/L	0.35–5.5uIU/L
FT4	1.12 ng/dL	0.80–1.80 ng/dL
Testosterone	**435.80 ng/dL(↑)**	2–8 ng/dL
FSH	**4.69 IU/L (↑)**	≤3 IU/L
LH	**2.36 IU/L (↑)**	0.02–0.3 IU/L
17-OH progesterone	1.90 ng/mL	Male: 0.5–2.1 ng/mL

as both his gonadarche and adrenarche were well advanced at presentation, it was difficult to draw such conclusions. His skeletal abnormality and skin lesion could be explained by both NF-1 and MCA syndrome. In MCA syndrome, the café au lait macules are irregularly bordered and usually do not cross the midline, although they may. They are often located on the same side as the main bone lesions and show a segmental distribution. It is often associated with hyperfunctioning multiple endocrine glands which were characteristically absent in this boy. Classically, the associated polyostotic fibrous dysplasia involves the skull base and long bones leading to bowing or other deformities. Our patient had a smooth bordered skin lesion crossing the midline suggestive of NF-1, and his skeletal lesions were in accordance with those of NF-1, not MCA syndrome.

His routine hematologic and biochemical parameters were normal. Hormonal investigations showed that he had serum total testosterone in adult-specific range with inappropriately elevated LH and FSH. Other hormonal reports were normal including thyroid and adrenal functions (Table 26.1).

Q2. What other investigations are required at this point?
  (a) GnRH stimulation test
  (b) HCG stimulation test
  (c) Pituitary MRI
  (d) USG of testis
  (e) Urinary ketosteroid

This boy had elevated gonadotropins as well as testosterone which indicated CPP, as a basal LH level >0.3 predicts with almost 100% certainty a pubertal response in the GnRH stimulation test. So, this stimulation test was unnecessary in this case. HCG stimulation test is commonly employed in the evaluation of 46XY DSD presenting with an empty scrotum or cryptorchidism to assess viable testis and also used in cases of testosterone, biosynthetic block, or receptor resistance. USG of the testis is recommended in peripheral precocity to detect such disorders like β-HCG-secreting tumor, Leydig cell tumor, adrenal rest tissue (TART), etc., which were the unlikely diagnoses in this boy. In congenital adrenal hyperplasia and testicular biosynthetic block, 17-urinary ketosteroid profiling is done to measure the precursor hormone level or its ratio with the active hormone. These were not indicated in our patient. As CPP is most likely to be organic in a young boy arising from such lesions as optic nerve glioma in NF1 or hypothalamic tumor, sellar/suprasellar MRI was the next essential investigation in our list.

MRI of the pituitary was done following adequate sedation of the boy. It showed a well-defined, rounded, isointense lesion (9.3 × 8.5 mm) in a hypothalamic area with the encasement of pituitary stalk, suggestive of hypothalamic hamartoma (Fig. 26.9).

Q3. How does hypothalamic hamartoma cause CPP?
- (a) Activation of GnRH pulse generator
- (b) Secretion of GnRH
- (c) Release of ectopic LH/FSH
- (d) Release of neurokinin B and dynorphin
- (e) Direct compression on the pituitary gonadotrope

Hypothalamic hamartoma is a developmental tumor containing GnRH neurosecretory cells that may directly liberate GnRH. It may also activate the GnRH pulse generator through the interconnected neuronal network. It does not secrete LH or FSH directly, and compression on gonadotropes would lead to hypogonadism rather than sexual precocity. Neurokinin B and dynorphin are the predominant neurotransmitters involved in the initiation of puberty. KNDY-Kisspeptin is stimulatory, whereas neurokinin B and dynorphin are inhibitory in nature. It is not yet documented that hypothalamic hamartoma secretes any of these factors.

This boy had a skin lesion typical of NF1. He had focal kyphosis and scoliosis which are the most commonly associated skeletal manifestations in NF-1. His father had a similar skin lesion, suggestive of familial inheritance. He underwent mutational testing for NF1 which was found to be positive. However, rare yet similar cases of co-occurrence of NF1 with hypothalamic hamartoma presenting with CPP have been reported in the literature. Currently, the boy is on GnRH agonist therapy and regular follow-up.

Discussion

Hypothalamic hamartomas are congenital malformations composed of a heterotopic mass of nervous tissue containing GnRH neurosecretory neurons, fiber bundles, and glial cells; they are frequently associated with CPP, which usually manifests before the age of 3 years. Hamartomas associated with CPP contain ectopic GnRH neurosecretory cells that are like GnRH-containing neurons in the medial basal hypothalamus [1]. These are postulated to elicit their effects via neurons that can produce GnRH within the tumor, through the ability to control neurons synaptically connected to GnRH neurons, or through neuronal networks that include GnRH neurons in the hypothalamic hamartoma itself or signaling-competent astrocytic and ependymoglial cells [2]. Precocious sexual development can be controlled by treatment with GnRH agonist. Hypothalamic hamartomas that cause ***CPP*** can be associated with laughing (gelastic), petit mal, or generalized tonic–clonic seizures; mental retardation; behavioral disturbances and dysmorphic syndromes beginning as early as the neonatal period. Surgical removal may lead to endocrinopathy,

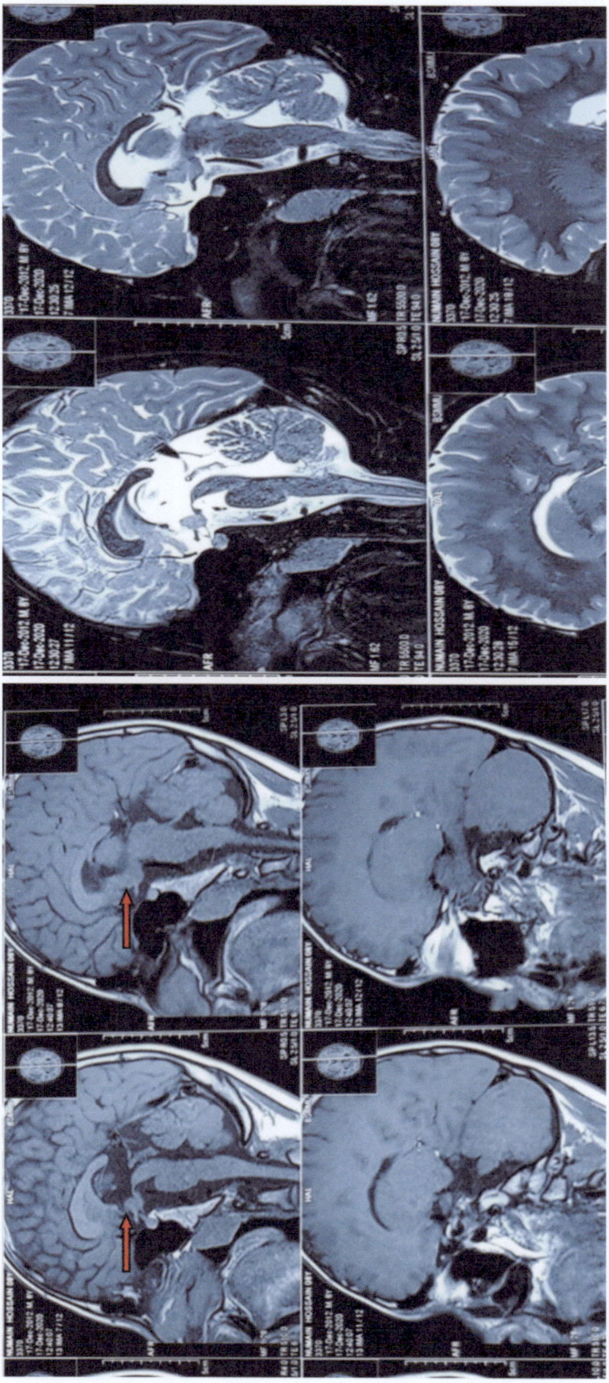

Fig. 26.9 Pituitary MRI showing a hypothalamic hamartoma

weight gain, and electrolyte imbalance [3]. It is only indicated in the case of intractable headache or seizure or progressive enlargement of the tumor. ***NF 1*** commonly is associated with optic nerve glioma that can lead to CPP by impinging on the GnRH pulse generator. It is manifested by various cutaneous as well as skeletal abnormalities, the commonest being kyphoscoliosis. The skin lesion in NF1 is smoothly bordered and widely distributed on both sides of the midline. On the other hand, MCA is characterized by unilateral, segmental skin lesions which are of irregular margin. It is often associated with multiple hormonal excesses, and it characteristically leads to peripheral sexual precocity by the activation of the GNAS1 gene. The co-occurrence of hypothalamic hamartoma and NF1 causing CPP is reported in only a handful of cases [4, 5]. The unique presentation of our patient illustrates that NF 1 need not be always associated with optic glioma for the activation of GnRH pulse generator and that coexistent hypothalamic hamartoma may produce the same picture.

Answer

Q1. (e)
Q2. (c)
Q3. (a), (b)

References

1. Kaplan SL, Grumbach MM. Pathogenesis of sexual precocity. In: Grumbach MM, Sizonenko PC, Aubert ML, editors. Control of the onset of puberty. Baltimore, MD: Williams & Wilkins; 1990. p. 620–60.
2. Grumbach MM, Kaplan SL. The neuroendocrinology of human puberty: an ontogenetic perspective. In: Grumbach MM, Sizonenko PC, Aubert ML, editors. Control of the onset of puberty. Baltimore, MD: Williams & Wilkins; 1990. p. 1–68.
3. Jung H, Neumaier PE, Hauffa BP, et al. Association of morphological characteristics with precocious puberty and/or gelastic seizures in hypothalamic hamartoma. J Clin Endocrinol Metab. 2003;88:4590–5.
4. Biswas K, Kapoor A, Jain S, Ammini AC. Hypothalamic hamartoma as a cause of precocious puberty in neurofibromatosis type 1: patient report. J Pediatr Endocrinol Metab. 2000;13(4) https://doi.org/10.1515/jpem.2000.13.4.443.
5. Bartolini E, Stagi S, Scalini P, Bianchi A, Ciccarone A, Mascalchi M. Central precocious puberty due to hypothalamic hamartoma in neurofibromatosis type. Hormones (Athens). 2016;15(1):144–6.

A 7-Year-Old Boy with Ambiguous Genitalia and Sexual Precocity

27

Introduction

This is a unique case scenario of a boy presented with genital ambiguity, followed by pubertal virilization resulting from a rare variety of CAH, namely 3-β-HSD type-2 deficiency. Biochemical evaluation for penoscrotal hypospadias and bilateral cryptorchidism along with growth acceleration and phallic enlargement revealed peripheral sexual precocity due to this unusual variety of CAH being diagnosed based on elevated precursor steroids and confirmed by genotyping. Interestingly, his karyotype was suggestive of mosaic KS that explained the presence of bilateral cryptorchidism. He underwent orchidopexy with steroid replacement that unmasked the hypergonadotropic hypogonadism attributable to KS.

Case Scenario

A 7-year-old boy was referred to us for the medical evaluation of penoscrotal hypospadias with bilateral cryptorchidism evident since birth. The child was the first issue of a consanguineous marriage (Fig. 27.1) born by normal vaginal delivery at a remote local hospital. The antenatal and peripartum periods were uneventful except for the baby's ambiguous genitalia that created much confusion on gender assignment but was eventually declared as a male on a clinical basis without any further evaluation. He then was brought up as a male with normal developmental milestones and satisfactory school performance (Fig. 27.1). However, at 7 years of age, he was noticed to have accelerated growth with the appearance of axillary and pubic hair with progressive blackening of the skin. Later on, his parents noticed the rapid phallic enlargement that brought him to medical attention. He had occasional nausea and diarrhea, however, denied having any weakness, fatigue, or postural dizziness.

S. Jahan, *Endocrinology Case Studies*,
https://doi.org/10.1007/978-981-95-2770-0_27

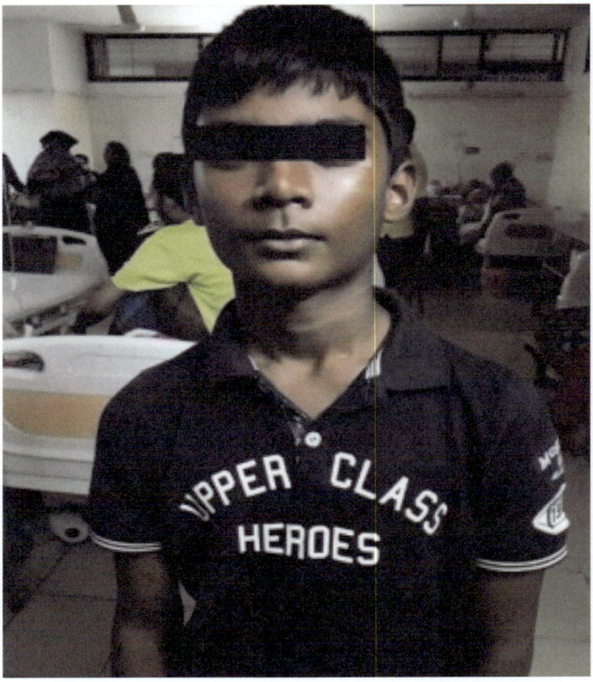

Fig. 27.1 Front profile of the boy showing prominent hyperpigmentation

Q1. What are the important points to be looked for in his history and physical examination?
 (a) Salt-wasting crisis
 (b) Headache and convulsion
 (c) Skeletal deformity
 (d) HTN/hypotension
 (e) Maternal drug ingestion during pregnancy

This boy with ambiguous genitalia progressed to pubertal virilization that could occur in various disorders of sexual differentiation. He represented a case of 46XY DSD who later on developed sexual precocity; the tempo of pubertal progression suggested peripheral rather than central etiology (pubic hair appearing earlier than phallic/testicular enlargement). Headache and convulsion, e.g., gelastic seizures, are characteristic of central precocity (CPP) resulting from hypothalamic hamartoma, optic nerve glioma in neurofibromatosis1 (NF1), dysgerminoma, etc. NF1 may present with CPP caused by optic nerve glioma, and it is commonly associated with such skeletal abnormalities as kyphoscoliosis, pectus ex-cavatum, etc. McCune Albright syndrome presents with a typical skeletal deformity in the form of polyostotic fibrous dysplasia and is characterized by hyperfunctioning endocrine glands with café-au-lait spots. It can sometimes lead to peripheral sexual precocity due to the activation of the GNAS1 gene in the gonads. Exposure to exogenous steroids during pregnancy can explain the genital ambiguity in this boy but not the pubertal virilization. There were certain features in history pointing toward congenital adrenal hyperplasia (CAH) including consanguinity, peripheral sexual precocity, progressive blackening, and occasional H/O nausea /diarrhea. Certain categories of 46 XY DSD and CAH are associated with HTN like 11-β-hydroxylase deficiency/17-α-hydroxylase deficiency, whereas others might be associated with hypotension, e.g., 21-α-hydroxylase deficiency depending on the type of precursor steroid overproduced.

There was no history of any salt-wasting crisis, hypertension/hypotension, maternal virilization during the antenatal period, intake of related drugs, or family history of a similar type of illness or any neonatal death. On examination, his height was just above the 95th percentile, body mass index was 12.71 kg/m^2, and the upper to lower segment ratio was 1.05. He had generalized hyperpigmentation, which was more prominent in the palmar creases, knuckles, oral mucosa, extensor surfaces, and on pressure areas (Fig. 27.1). Though he was normotensive for age, he showed evidence of postural hypotension. The rest of the general and systemic examinations including fundoscopy and other vital signs were normal. Sexual maturity rating was consistent with pubic hair—Tanner stage IV, stretched penile length (SPL) of 4 cm, and external masculinization score (EMS) was −3. He had penoscrotal hypospadias, chordae, and his scrotal skin was darkened and rugosed but with incomplete fusion, and the testes were impalpable within the sac or inguinal region (Fig. 27.2).

Routine blood test results including liver and renal functions, thyroid hormone profile, and blood sugar were all normal. His radiologic bone age was advanced, between 16 and 17 years. He had normo-natremia with marginal hyperkalemia

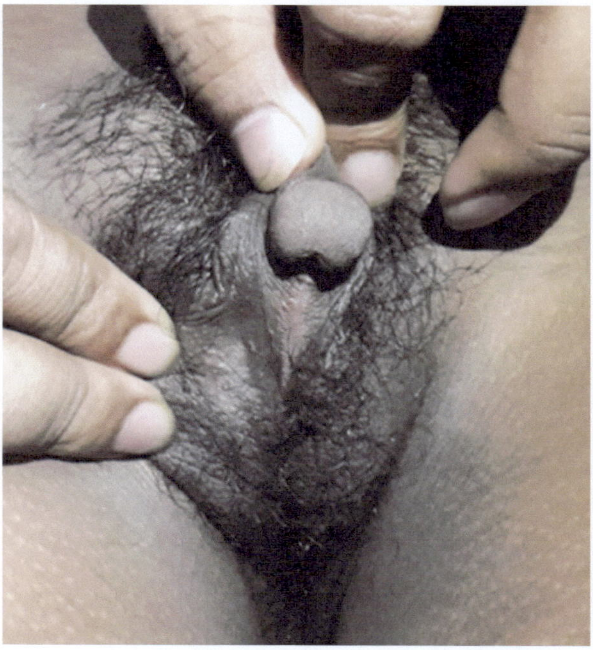

Fig. 27.2 Ambiguity of the external genitalia. (**The photograph was taken with the permission of the parent)

Table 27.1 Pretreatment hormonal report of the boy showing high testosterone (both basal and poststimulation) and low gonadotropins with elevated precursor steroids

Hormonal parameter	Test result	Reference value
Serum LH	0.19 mIU/mL	≤ 0.44 mIU/mL
Serum FSH	3.83 IU/L	< 3 IU/L
Serum testosterone	**276 ng/dL**	≤ 42 ng/dL
Serum testosterone (following hCG)	**561.63 ng/dL**	NA
Serum estradiol	11.8 pg/mL	≤ 4 pg/mL
Basal cortisol	**129 nmol/L**	138–690 nmol/L
Plasma ACTH	**269 pg/mL**	0.00–46 pg/mL
Plasma 17-OH-progesterone	2.15 ng/mL	0.5–2.1 ng/mL
Plasma 17-OH-pregnenolone	**2097 ng/dL**	≤ 72 ng/dL
Plasma DHEA-S	**331.80 μg/dL**	< 186 μg/dL
Plasma androstenedione	34 ng/dL	6–115 ng/dL
Plasma dihydrotestosterone (DHT)	28 ng/dL 70 ng/dL (post-stimulation)	14 to 77 ng/dL (baseline) NA
Plasma renin	6.2 ng/dL	0.4–3 ng/dL
Plasma aldosterone	8.5 ng/dL	3–40 ng/dL

(Na—134, K—5.7 mmol/l). Hormonal assay (Table 27.1) showed low serum LH and FSH with inappropriately high age-matched serum total testosterone levels. Serum basal cortisol was low with elevated adrenocorticotrophic hormone (ACTH); he had high normal 17(OH) progesterone and disproportionately elevated dehydroepiandrosterone sulfate (DHEA-S). High baseline ACTH and 17(OH) progesterone obviated the need for an ACTH stimulation test. An HCG stimulation test exhibited an elevated testosterone level from baseline to 561.63 ng/dL, confirming the presence of intra-abdominal viable testis (Table 27.1).

Q2. What is the possible reason of pubertal virilization in this boy with ambiguous genitalia?
 (a) 5-α-reductase type 2 deficiency
 (b) Partial androgen insensitivity syndrome (PAIS)
 (c) Ovotesticular DSD
 (d) 17-β-HSD type 3 deficiency
 (e) CAH

46 XY DSD can result from a diverse condition including gonadal dysgenesis, steroid biosynthetic defect, defective conversion of testosterone into dihydrotestosterone (DHT), androgen receptor resistance, etc. Some of these disorders are characterized by pubertal virilization which might even change the previous female gender assignment given at birth. 5-α-reductase type 2 deficiency is such an enzyme defect leading to the failure of conversion of testosterone to DHT. In this case, the presentation is with ambiguous genitalia, followed by some degree of virilization at puberty due to unaffected activity of 5-α-reductase type 1 derived from the skin. On

the other hand, PAIS also presents with microphallus/hypospadias, and resistance to androgen is partially overcome in puberty leading to a variable amount of pubarche or phallic enlargement. 17-β-HSD type 3 deficiency is characterized by the absence of the named enzyme activity necessary in the terminal step of sex steroid synthesis converting androstenedione to testosterone. It might have a similar presentation owing to extragonadal conversion of androstenedione to testosterone sufficient for pubertal virilization. This boy had pronounced hyperpigmentation, postural hypotension with hyperkalemia, low cortisol and aldosterone with high ACTH, low gonadotropin, and high testosterone level. This biochemical picture is consistent with congenital adrenal hyperplasia with a proximal block in both glucocorticoid and mineralocorticoid synthesis pathways. None of the other differentials above affect cortisol/mineralocorticoid synthesis, so they were less likely in this case. Considering the sex hormone profile, his testosterone: androstenedione ratio was >8 which would typically be <0.8 in the case of 17-β-HSD type 3 deficiency. On the other hand, his testosterone: DHT ratio in the post-HCG stimulation test was normal (8:1), which is expected to be >10:1 in suspected PAIS. In PAIS, elevated testosterone should be associated with high gonadotropin rather than low as in this case. Ultimately, his clinical and biochemical phenotype was suggestive of a specific category of CAH presenting with ambiguous genitalia leading to virilization in later years.

He underwent further investigations like computed tomography (CT) scan of the abdomen, which showed a soft tissue density structure, measuring about (15 × 13) mm at the right side of the pelvic cavity in the region of inguinal canal likely representing right testis; however, left testis could not be delineated, and both the adrenals were mildly enlarged, (25 × 7.9) mm on the right and (26 × 6.7) mm on the left side with uniform postcontrast enhancement. His karyotyping report was quite unexpected: mosaic Klinefelter syndrome—47XXY/46XX (3:1).

Q3. What is the probable variety of CAH this boy had?
   (a) 21-α-hydroxylase deficiency
   (b) 11-β-hydroxylase deficiency
   (c) 3-β-HSD type 2 deficiency
   (d) 17-α-hydroxylase deficiency
   (e) 17-β-HSD type 3 deficiency

21-α-hydroxylase deficiency is the commonest variety of CAH presenting with peripheral sexual precocity in a phenotypic boy, which does not cause genital ambiguity/microphallus or hypospadias in a male child. Similarly, 11-β-hydroxylase deficiency is associated with hypokalemic HTN, whereas the former one may have hypotension with hyperkalemia as in our case. 17-α-hydroxylase deficiency leads to hypergonadotropic hypogonadism presenting with female phenotype in a genotypic male. It is often associated with hypokalemic HTN due to increased production of deoxycorticosterone and 11-deoxycortisol. None of the abovementioned varieties lead to pubertal virilization except 17-β-HSD type 3 deficiency and 3-β-HSD type 2 deficiency. The latter one was the strongest possibility here as it presents with genital ambiguity, hyperkalemic hypotension, hyperpigmentation, occasional vomiting/diarrhea resulting from elevated DHEA-S, low mineralocorticoid, high ACTH, and low cortisol, respectively. It often leads to pubertal virilization because of HSD3B1-mediated peripheral conversion of Δ5-C19-steroids to Δ4-C19-steroids. On the other hand, 17-β-HSD type 3 deficiency does not affect glucocorticoid and mineralocorticoid pathways, so it was an unlikely diagnosis in this boy.

Q4. What are the definitive tests for diagnosing the variety of CAH in this boy?
- (a) Plasma 17(OH) progesterone level
- (b) Plasma DOC and 11-deoxycortisol level
- (c) 17(OH) pregnenolone level
- (d) Genetic mutational analysis for 3-β-HSD2
- (e) Urinary Δ4-C19-steroids profile

CAH is categorized on the basis of the elevated level of the metabolite proximal to the enzymatic block, e.g., plasma 17(OH) progesterone or plasma DOC and 11-deoxycortisol level in 21-α-hydroxylase and 11-β-hydroxylase deficiency, respectively. 3-β-HSD2 deficiency is a rare variety of CAH affecting both adrenal and gonadal steroid synthesis. It is characterized by the failure of conversion of Δ5-C19-steroids to Δ4-C19-steroids {e.g., 17(OH) pregnenolone to 17(OH) progesterone}. In our case, the 17(OH) pregnenolone level was markedly elevated (usually >100 nmol/L is suggestive) (Table 27.1). The increased ratio of Δ5-C19-steroids to Δ4-C19-steroids in urine also supports this; however, it was not done in this case because of the lack of local availability. The final diagnosis relies on genetic mutational analysis for 3-β-HSD2; sequencing of HSD3B2 gene was carried out that showed the two homozygous missense mutations V299I (GTA > ATA) and S309T (TCC > ACC) and one heterozygous mutation Q311R (CAA > CGA) (Fig. 27.3).

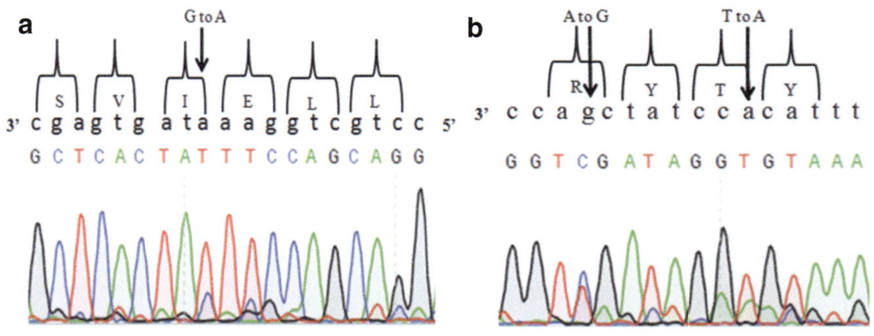

Fig. 27.3 Partial nucleotide sequences of *HSD3B2* gene showing in black arrows the homozygous G > A transversion (**a**) and heterozygous A > G transversion and T > A transition (**b**) in antisense strands. Upper panels in small letters show the complementary sequences of the nucleotides, and three-letter blocks show the respective amino acids

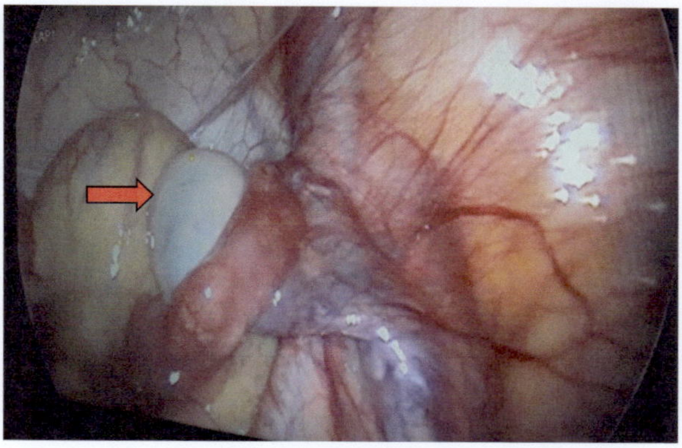

Fig. 27.4 Intra-abdominal testis on the right side during exploratory laparotomy

Table 27.2 Hormonal parameters following treatment of the patient unmasking hypergonadotropic hypogonadism

Hormonal parameter	After treatment	Before treatment
Serum LH	3.05 IU/L	0.19 mIU/mL
Serum FSH	13.96 IU/L	3.83 IU/L
Serum testosterone	107.50 ng/dL	276 ng/dL
Plasma ACTH	7.19 pg/mL	269 pg/mL
DHEA-S	26.70 μg/dL	331.80 μg/dL

The patient was discharged on tab. Hydrocortisone 5 mg in the morning and 10 mg at night and tab. Fludrocortisone 0.1 mg daily along with necessary advice. One month later, the urologist performed orchidopexy of the right testis, whereas the left testis found to be rudimentary was sacrificed (Fig. 27.4). After 3 months, he had subjective improvement in terms of remission of fatigue, nausea, and diarrhea; his skin became less pigmented, though he developed mild tender gynecomastia. At subsequent follow-up at 6 months, the gynecomastia resolved, and the hormonal profile showed suppression of ACTH and adrenal androgens with the biochemical unmasking of hypergonadotropic hypogonadism due to KS (Table 27.2). At present, correction of hypospadias, as well as genetic screening of the rest of the family members, is awaited. In the future, he will probably require androgen replacement when his hypogonadal features become more evident.

Discussion

CAH due to 3-β-HSD deficiency is a rare autosomal recessive disorder that results from mutations within the HSD3B2 gene encoding 3-β-HSD type 2 enzymes that affect both adrenal and gonadal steroid production. The enzyme HSD3β2 catalyzes 3-beta-dehydrogenation and isomerization of the double bond of the steroid B ring to the steroid A ring, converting Δ5 to Δ4 steroids such as pregnenolone to progesterone (mineralocorticoid pathway), 17-α-hydroxypregnenolone to 17-α-hydroxyprogesterone (glucocorticoid pathway), dehydroepiandrosterone (DHEA) to androstenedione and androstenediol to testosterone (sex steroid pathway). Therefore, the absence of this enzyme impairs all these Δ4 steroid production, while low levels of cortisol result in increased ACTH stimulation of steroids prior to the 3-β-HSD step, producing increased accumulation and secretion of Δ5 3-β-hydroxy steroids: pregnenolone, 17-α-hydroxypregnenolone, and DHEA. Adrenal insufficiency generally occurs secondary to aldosterone and cortisol deficiency. 46 XY individuals with severe defects may present with ambiguous genitalia, hypospadias, severe adrenal insufficiency in infancy, poor virilization at puberty, and gynecomastia but milder forms with penoscrotal hypospadias and premature adrenarche. Indeed, this differentiation is not straightforward, and impaired male sexual differentiation correlates poorly with salt-wasting forms; moreover, identical mutations have also been found in both forms [1]. Virilization or spontaneous puberty has been reported in patients secondary to either direct effects of DHEA or to sufficient conversion of DHEA to testosterone via peripheral type I 3-β-HSD isoenzyme [1]. The simultaneous occurrence of CAH and ***Klinefelter syndrome (KS)*** is extremely rare. The interest in describing this case lies in the exceptionality of the concurrent occurrence of CAH because of 3-β-HSD deficiency and mosaic KS 47, XXY/46, XX (3:1). This provides clinical evidence of the coexistence of two diseases, the first of which causes androgen excess and the second one androgen deficiency. Most patients with 3-β-HSD deficiency require treatment with mineralocorticoids, glucocorticoids, and androgen replacement. Suppression of adrenal sex steroid production by glucocorticoids might lead to the unmasking of hypogonadism resulting from KS which probably would require future androgen replacement in the boy of the above case scenario.

Answer Key

Q1. (a), (c), (d), (e)
Q2. (e)
Q3. (c)
Q4. (c), (d)

Reference

1. Achermann JC, Hughes IA. Pediatric disorders of sex development. In: Melmed S, Polonsky KS, Larsen PR, Kronenberg HM, editors. Williams textbook of endocrinology. Philadelphia, PA: Elsevier; 2016. p. 893–963.

Index

S. Jahan, *Endocrinology Case Studies*,
https://doi.org/10.1007/978-981-95-2770-0

MIX
Papier aus verantwortungsvollen Quellen
Paper from responsible sources
FSC® C105338

If you have any concerns about our products,
you can contact us on
ProductSafety@springernature.com

In case Publisher is established outside the EU,
the EU authorized representative is:
Springer Nature Customer Service Center GmbH
Europaplatz 3, 69115 Heidelberg, Germany

Printed by Libri Plureos GmbH
in Hamburg, Germany